SWINE PRODUCTION AND MANAGEMENT

Among the different livestock enterprises, pig farming has transformed itself into the most profitable livestock enterprise. They produce more live weight gain from a given quantity of feed than any other class of meat producing animals. They are prolific with shorter generation interval and converts vegetable and slaughter waste into valuable protein for human consumption. Pork is nutritious with high fat and lower water content and has got better energy value than that of other meats. It is rich in vitamins like thiamine, niacin and riboflavin.

The book swine production and management elaborately covers all topics of swine management like breeding, feeding, housing, health management and pork production technology. It can be easily read and understood as the language is simple and lucid. The topics are discussed point wise for an examination perspective. The book is well supported by a large number of illustrations and tables which makes the understanding of the text very simple and easy. It will be very useful for all students as well as professionals.

It also provides scientific principles lucidly so as to translate the application of these principles to suit the varied situations prevalent in developing countries. An emerging hot topic 'Bio-fuel production from pork fat' is discussed elaborately as 'Chapter-8' for the first time in this book.

Dr. John Abraham Assistant Prof. [MVSc., Ph.D., MBA., MSc (Psy)] Graduated in the year 1999 from the College of Veterinary and Animal Sciences, Mannuthy, Thrissur, followed by Post Graduation from the Indian Veterinary Research Institute, Izatnagar, Bareilly. Took M.B.A form Tamil Nadu Agricultural University in the year 2010. Bagged 1st rank and 4 gold medals for Ph.D, awarded by Tamil Nadu Veterinary and Animal Sciences University in the year 2012. Invented the technology to produce biodiesel from dead poultry (under patenting), which was awarded the prestigious 'DST-Lockheed Martin-India Innovation Growth Programme Award-2016'. Awarded the 'Millennium Alliance Award-2016' for the invention "Low cost manual milking machine".

SWINE PRODUCTION AND MANAGEMENT

By

Dr. John Abraham

Department of L.P.M,
College of Veterinary and Animal Sciences,
Pookot, Wayanad-673576, Kerala

CRC Press
Taylor & Francis Group
Boca Raton London New York

CRC Press is an imprint of the
Taylor & Francis Group, an **informa** business

NARENDRA PUBLISHING HOUSE
DELHI (INDIA)

CRC Press
Taylor & Francis Group
6000 Broken Sound Parkway NW, Suite 300
Boca Raton, FL 33487-2742

CRC Press is an imprint of the Taylor & Francis Group, an informa business

International Standard Book Number-13: 978-0-367-49085-0 (Hardback)

Print edition not for sale in South Asia (India, Sri Lanka, Nepal, Bangladesh, Pakistan or Bhutan)

Library of Congress Cataloging-in-Publication Data
A catalog record has been requested

Visit the Taylor & Francis Web site at
http://www.taylorandfrancis.com

and the CRC Press Web site at
http://www.crcpress.com

Contents

Preface

With the world's population predicted to reach nine billion by 2050, we face the challenge to ensure the supply of affordable nutritious food, without decimating the earth's natural resources. Poor diet stunts the growth of 162 million children every year, 97 per cent of them in developing countries. A doubling of the global food demand projected for the next 50 years pose huge challenges. The challenges that are emerging, ranging from climatic change to unforeseen disease and pests. These necessitate understanding the challenges; possible direction of future farming, processing, marketing and drawing conclusions on future crop livestock interaction.

Among the different livestock enterprises, pig farming has transformed itself into the most profitable livestock enterprise. They produce more live weight gain from a given quantity of feed than any other class of meat producing animals. They are prolific with shorter generation interval and converts vegetable and slaughter waste into valuable protein for human consumption. Pork is nutritious with high fat and lower water content and has got better energy value than that of other meats. It is rich in vitamins like thiamine, niacin and riboflavin. Pig farming provides quick returns, since the marketable weight of fatteners can be achieved with in a period of 6-8 months. The dietary diversification studies carried out in 2018 reveals that, as income rise, the demand tends to shift towards protein rich foods of animal origin. Even a tiny increase in per capita consumption has the potential to translate into a significant rise in market volumes.

There is a growing demand for pork due to increasing per capita income and urbanisation.

Pig farming is still a much neglected livestock sector in developing countries. Animals are reared as scavengers in small holdings. This is gradually getting transformed with increase in herd size and implementation of scientific management practices into profitable enterprises. There are lacunae in knowledge dissemination and implementation of scientific management practices. In this context, this books is written to provide comprehensive knowledge and information on all aspects of swine production. It also provides scientific principles lucidly so as to translate the application of these principles to suit the varied situations prevalent in developing countries. An emerging hot topic 'Bio-fuel production from pork fat' is discussed elaborately as 'Chapter-8' for the first time in this book.

I hope that the students of Animal Science across contents will find this book interesting and useful.

John Abraham

2019-06-26

CHAPTER 1

SWINE PRODUCTION

Swine husbandry has enormous potential and prospects because of its excellent attributes such as high fecundity (ability to produce more number of ova regardless whether it gets fertilised or not), prolificacy (ability to produce more number of live off-springs in one farrowing), short generation interval, quick body weight gain, high feed conversion efficiency, and excellent dressing percentage compared to all other livestock.

The recommended meat consumption level is 34 g/head/day, but the availability of animal protein per capita per day to an average Indian is only 12.6 g/head/day. There is a big gap between the demand and supply. Due to the prohibition of cattle slaughter in various states, the increasing demand for meat of our growing population cannot be met by ruminants alone, which have the drawback of relatively low rate of weight gain and large generation interval. In this context, swine husbandry has great prospects and potential to provide protein rich food to millions of people.

Among livestock, pigs are the only species reared exclusively for meat purpose and it fits in extremely well with the mixed farming system and is complementary to intensive crop production operation. It's unique ability to efficiently convert swill (waste food), tankage, (slaughter waste) and agricultural by-products into food source of high quality animal protein and other nutrients, make it the species of choice for the most profitable livestock enterprise and for alleviating malnutrition among people who lives below the poverty line. It also provides gainful opportunities for unemployed rural and urban population. Thus piggery have now carved a niche for itself in North and North-Eastern states and is also becoming popular in most of the Southern regions of the country.

Taxonomical Classification of Swine

Kingdom	-	Animalia
Phylum	-	Chordata
Sub Phylum	-	Vertebrata
Class	-	Mammalia
Sub Class	-	Utheria
Order	-	Artiodactyla
Family	-	Suidae/Porcidae
Genus	-	Sus (*Linn*)
Species	-	*Sus scrofa*
		Sus domesticus
		Sus vittatus
Sub Species	-	Philipinensis, cristatus, strotzai.

Chordata: One of the 21 phyla of the animal kingdom in which there is either a back bone (In the vertebrates) or the rudiment of a back bone-the notochord.

Mammalia: They are warm blooded hairy animals that produce their young ones alive and suckle them for a variable period, from the secretion of the mammary gland.

Artiodactyla: Even toed hoofed mammals.

Family-Suidae: The family of non-ruminant, artiodactyl ungulates, consisting of wild and domestic swine but, in modern classifications excludes the peccaries.

Genus-Sus: The typical genus of swine, formerly comprehensive but, now restricted to the European wild boar and its allies, with the domestic breeds derived from them.

Species-Sus scrofa: It is the wild hog of continental Europe from which most of the domestic swine have been derived.

Sus vitatus: It was the chief, if not the only race or species of the East-Indian pig that contributed to present-day domestic swine.

Origin of the Domestic Pigs

Pigs were probably domesticated by 490 B.C in China and by 800 B.C in England. It appears that the ancestor of the common domestic pig belonged to a single wild species, *Sus scrofa,* formerly extant throughout Eurasia and also found in North-Africa and Nile Valley. About 25 sub species of *Sus scrofa* have been described (Clutton-Brock, 1981) and it is assumed that they evolved by adaptations to localised environment. In addition, there are four other sub-species and one related animal (National Research Council, 1998) found only in the South-East Asia. These are:

1. *Sus barbatus* (*the bearded pig*) A quite large (<150 kg) animal with red or black coat found in Philippines, the island of Borneo and Sumatra in Indonesia and the Malay peninsula. It had not been domesticated, but it hybridises with *S. scrofa*, both sexes being fertile.
2. *Sus celebesis* (*the Sulawesi Warty pig*) A medium sized (<70 kg) animal, with a red/brown coat. It is found in Sulawesi and other Eastern islands of Indonesia. It has been domesticated on the island of Roti and is one of the ancestors of domesticated New Guinea pigs.
3. *Sus verrucosus* (*the Javanese Warty pig*) A large (<120 kg) animal with yellow, red or black hair, found in Java and some adjacent island of Indonesia. It has not been domesticated, but it hybridise with *S. scrofa*. It is an endangered species.

4. *Sus salvanius* (*the pygmy hog*) A very small (<10 kg) animal once widely distributed in the foot hills of the Himalayas, now only known definitely to occur in the Manas National park in Assam, India.
5. *Babyrousa babyrussa (the babirusa)* A pig like animal, living in the forest of Sulawesi and adjacent island of Indonesia.

Domestic pigs have been derived from one or other of the many subspecies of *S. scrofa*. Three sub-species may have been of particular importance: *Sus scrofa scrofa*, the Eurasian wild pig, *Sus scrofa cristatus*, the Indian jungle pig and *Sus scrofa vittatus* a wild pig of South East Asia.

Body parts of the pig

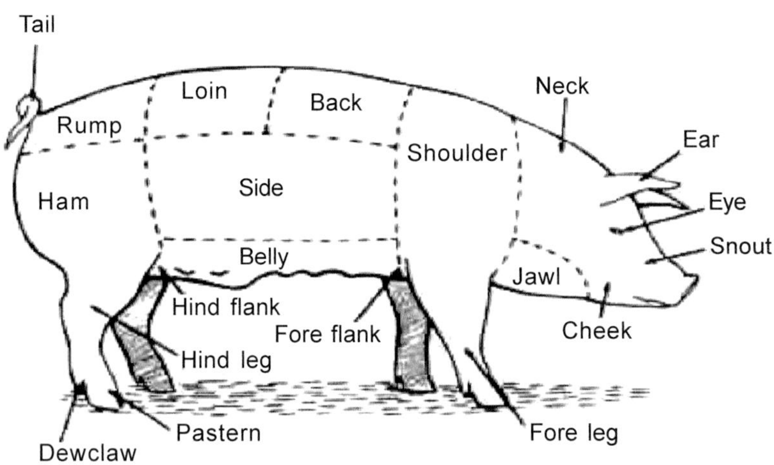

FIG.1.1 BODY PARTS OF THE PIG

The knowledge of body parts of swine is useful for the study of breed characteristics, judging the animal, locating external abnormalities on the body, writing reports of treatment and dressing of the parts and for proper classification of market hogs.

BREEDS OF PIGS

Throughout the history, breeds were developed and maintained either by geographical isolation or by the efforts of man by years of breeding and selection for a particular trait to meet the demand for a desired type. Worldwide, there are 355 recognised breeds of pig. In the case of pigs, over the years, the demand has moved from a lard-type pig (carcass high in fat) to the present bacon type pigs (carcass high in meat compared to lard).

Indian Breeds of Pig

FIG. 1.2 THE INDIAN WILD BOAR (*Sus scrofa cristatus*)
Source: http://wildindia.com

The Indian wild boar is commonly found in jungles and forest. On Himalayas it is present up to an elevation of 4.5 km. The adult animal measures about 1.5 m in length from nose to vent and weigh up to 136 kg. The wild pig has a long snout, short ribs and long legs. Males are larger than females. Colour of the animal is rusty grey when young which gradually transforms to dark chestnut brown with its hair tinged grey at the extremities as its age advances. The wild pig is distinctive in possessing a sparse coat and a full crest of mane of black bristles running from the nape down the back. The tusks are well developed in males, both the upper and lower tusks curving outwards and projecting from the mouth. The wild pig is extremely active and when provoked, may also attack human beings.

Wild pigs are highly prolific and breed in all seasons. To provide shelter for the young ones before farrowing, the females dig the ground and make bedding of grass and straw. Piglets (4-6) are born in a litter after a period of four months of gestation. They move in herds of 10-20. Being omnivorous, they live on plants, roots, tubers, insects, offal, carrion *etc.* No animal is more destructive to crop than the wild pig in cultivated areas. It's meat is, however very delicious.

The Pigmy hog (*Sus salvanius*)

FIG.1.3 THE PIGMY HOG
Source: http://professorpaulanimalencyclopedia.blog.spot.com

The pigmy hog is found to inhabit the dense moist forest (generally *Sal*) at the base of the Himalayas in Sikkim, Nepal, Bhutan and Assam. It is nocturnal in habit and prefers to remain in high grasses and therefore, is rarely seen. It lives in herds of 5-20. The animal measures about 32 cm over the shoulder and 66 cm from snout to rump and weighs 7.7 kg. Its colour is brown and black. No distinct crest is present.

The hairs on the hind parts of the neck and middle of back are rather long whereas those of ears are small. It's habits are similar to those of the wild boar.

Ankamali Pigs

FIG 1.4 ANKAMALI PIG
Source: National Bureau of Animal Genetic Resources, Karnal

Ankamali pigs evolved from domestication of *Sus scrofa andamansis* and are found in Kerala, Karnataka, Tamil Nadu and Maharashtra. They are medium sized, unicoloured rusty gray or multicoloured black with white patches. The body measures from 90 to 100 cm from snout to rump. The adult weighs 40-70 kg. The tail is black in colour and reaches nearly to the hock.

FIG. 1.5 WILD PIG AT THEKKADY AND MUTHANGA WILD LIFE SANCTUARY, KERALA

Recognised Indian Breeds of Pigs

The National Bureau of Animal Genetic Resource (NBAGR), Karnal has registered seven breeds of Indian pigs. They are Ghungroo from West Bengal, Niang Megha from Meghalaya, Agonda Goan from Goa, Tenyi Vo from Nagaland, Nicobari from Andaman & Nicobar, Doom from Assam and Zovawk from Mizoram.

Ghungroo pigs

FIG.1.6 GHUNGROO PIG
Source: National Bureau of Animal Genetic Resources, Karnal

Ghungroo is an indigenous breed of pig first reported from North Bengal (Darjeeling, Jalpaipuri, cochbehar and North and South Dinajpur. Ghungroo are mostly black coloured with typical Bull dog face appearance, with a litter size of 6-12 piglets, individually weighing about 1.0 kg at birth and 7.0 – 10.0 kg at weaning. Both sexes are very much docile and easy to handle. At the National Research Centre on Pig, Rani, Guwahati, Ghungroo pigs are being maintained under intensive system of rearing. Ghungroo pigs are black in colour with broad, flattened face and short, upwardly curved snout and large, pendulous ears. The body top line is straight with flat type belly. Bristles are thick, coarse and long. A female Ghungroo pig bears 10-12, teats with the ability to give birth to an average of 9 piglets at first farrowing.

Niang Megha

FIG.1.7 NIANG MEGHA
Source: National Bureau of Animal Genetic Resource

Niang Megha is a pig breed from Meghalaya (Garo, Khasi and Jaintia hills) reared for pork and bristle. The animal is mostly black with star shaped white patches at forehead and sometimes on hock and feet. It has long tapering snout and white at nostrils. It is pot-bellied and ears are small and erect.

Agonda Goan

FIG.1.8 AGONDA GOAN
Source: National Bureau of Animal Genetic Resource

It is a large sized local pig of Goa. It originated from southern taluka of Canacona. Can grow up to 160 kg. Mostly black coloured with some white patches on legs and face. It has short snout and rough bristles, black and gray in colour. People prefer this pig for sausage making and it is well adapted to local costal environment. Adapt to hot and humid climate; higher disease resistance and stress tolerant. It has high dressing percentage up to 80%.

Tenyi Vo

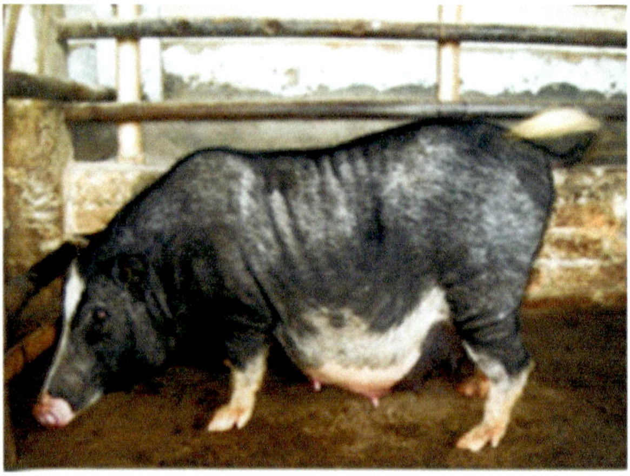

FIG.1.9 TENYI VO
Source: National Bureau of Animal Genetic Resource

It is a native Naga pig. It is found in the areas of the districts of Chakesang, Mao, Tuensang and Angami in Nagaland. This breed is a small pig with white hooves (sometimes nicknamed "the pig with white stockings"). Their bodies are black in color and hairy, with a potbelly. Tenyi Vo pigs have a strong and long tapering snout, small erect ears and bright alert eyes. It is predominantly black in colour with white markings on the forehead, flanks and legs. The tail is long and usually ends with a white switch. Adult pigs weigh from 35 to 50 kg. The estimated population is about 60,000 to 70,000.

Nicobari- Pigs of Andaman Nicobar Islands- (*Sus scrofa andamansis*)

FIG.1.10 NICOBARI
Source: National Bureau of Animal Genetic Resource

Exclusively reared by Nicobari tribes for meat purpose under free range or semi-intensive system. Feeds on coconut, root crops, fish waste and crabs. It is a dark grey coloured wild pig found in the Andaman and Nicobar Islands. Ears are straight, legs are greyish with white hooves. Adult weighs around 80-170 kg. By 10-12 months of age, they attain maturity. Their average litter size is 5-6 piglets. This breed is well adapted to the local environment of Nicobar and can survive with a very low level of management. At present, the Nicobari-indigenous pig breed is under the threat of extinction.

Doom

The Doom pig is a native variety of pig found in pure form in its breeding tract in Assam (Dhubri, Bongaigaon and some parts of Kokrajhar district). Body is comparatively larger than Assam local breeds. A medium sized animal with black colour. Short pointed snout. Small horizontally spreading ears. Erect mane spreading from forehead to back of shoulder, extending up to lumber regions. The snout is short and concave. Adult female weighs 36-50 kg but boar attains up to 70 kg.

FIG.1.11 DOOM
Source: National Bureau of Animal Genetic Resource

Zovawk

Zovak Male

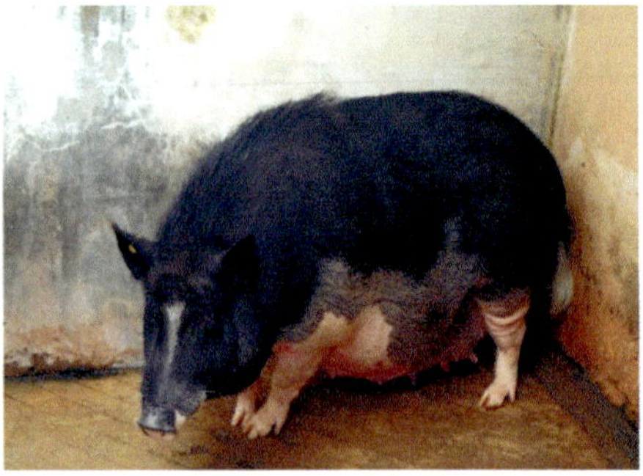

Zovak Female

FIG.1.12 ZOVAWK

Zovawk pig is distributed in Mizoram with an estimated population of about 39,000. It is reared by Mizo community for pork. Animals are black with white spots on fore head, white patches on belly and white stockings. Erect ears, concave snout, potbelly and concave top line and long bristles on mid-line. Average body weight is 54 kg for males and 59 kg for females.

Desi Pigs

FIG. 1.13 DESI PIGS

These pigs differ in their characteristics and colour from region to region depending on the topography and climatic condition. Different colour patterns are found, *viz.*, unicoloured rusty grey, brown, black, and even an admixture of these colours. They vary in size and appearance. They are commonly found in U.P, Bihar, Madhya Pradesh, Jharkhand, Uttaranchal and Punjab.

EXOTIC BREEDS OF SWINE

White Breeds

Large White Yorkshire

FIG 1.14 LARGE WHITE YORKSHIRE

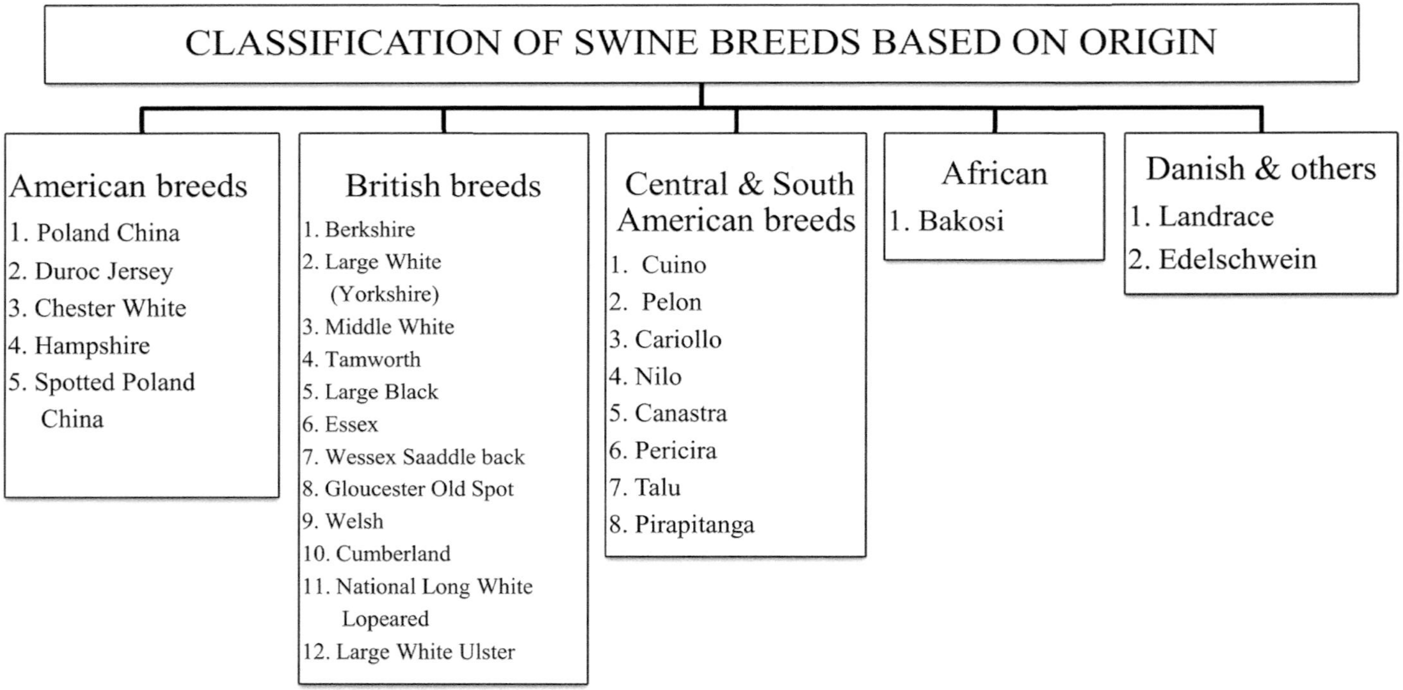
CLASSIFICATION OF SWINE BREEDS BASED ON ORIGIN
American breeds
1. Poland China
2. Duroc Jersey
3. Chester White
4. Hampshire
5. Spotted Poland China
British breeds
1. Berkshire
2. Large White (Yorkshire)
3. Middle White
4. Tamworth
5. Large Black
6. Essex
7. Wessex Saaddle back
8. Gloucester Old Spot
9. Welsh
10. Cumberland
11. National Long White Lopeared
12. Large White Ulster
Central & South American breeds
1. Cuino
2. Pelon
3. Cariollo
4. Nilo
5. Canastra
6. Pericira
7. Talu
8. Pirapitanga
African
1. Bakosi
Danish & others
1. Landrace
2. Edelschwein

Classification of swine based on utility

Lard type

1. Duroc
2. Poland China
3. Chester White
4. Berkshire
5. Hampshire
6. Spotted Poland China

Bacon type

1. Tamworth
2. Yorkshire

Classification of Swine based on Colour

Colour	Breeds
1) White breeds	Yorkshire, Chester white and Landrace
2) Black with some white	Hampshire, Berkshire and Poland china
3) Black and white breeds	Spots
4) Black breeds	Large Black
5) Red breeds	Duroc and Tamworth
6) Red with some white	Hereford

The Large White Yorkshire is a popular English bacon breed which had its origin nearly a century ago in Yorkshire and neighbouring places in northern England. The Yorkshire carries the title of "The mother breed" because the Yorkshire sow is known for large litters and good mothering ability. The Yorkshire is a good feed converter and yields a carcass of high dressing per cent. Three types of Yorkshires, the large white, middle white and small white are reported.

Colour: Entirely white in colour. Black pigment spots called "*freckles*" do not constitute a defect though they are not preferred by breeders.

Head: Moderately long, face slightly dished, snout broad and wide between the nostrils and the ears.

Ears: The ears are upright.

Neck: Long, fine and proportionately full to shoulder with wide and deep chest.

Back: Long level and wide from neck to rump.

Legs: Straight and well-set, level with the outside of the body with flat bone.

Skin: Fine, white, free from wrinkles.

Live weight: Mature boars weigh from about 300 to 450 Kg, while average sow weigh from 250 to 350 Kg.

Carcass quality: Possess necessary length, gives a first grade bacon, good ham and comparatively light fore quarter.

Disqualification: Swirls (A swirl is hairs growing in a circular pattern from the roots) on the upper third of the body, hair other than white and blind or inverted teats. Other disqualifications include less than six teats on a side, hernia and cryptorchidism.

This breed is known to do well under Indian condition. Used in two breed or three breed crosses.

Middle White Yorkshire

FIG. 1.15 MIDDLE WHITE YORKSHIRE
Provided by Michael von Luttwitz

This breed was evolved in Yorkshire of Northern England by crossing the Large white with smaller breeds of Yorkshire extraction. The breed is accepted as an excellent pork pig, reaching slaughter weight at early age and with a high percentage of lean meat to bone. The breed is hardy, grows rapidly, but is not as prolific as the Large white. In India, the breed has been extensively used for improving indigenous stock in rural areas.

Colour: White.

Head: Moderately short, straight jaw, short snout.

Neck: Fairly light, medium length, proportionately and evenly set on the shoulder.

Ears: The ears are upright.

Face: Upturned dished face, wide between ears.

Back: Long and level to root of the tail with well sprung ribs.

Legs: Straight and fairly short, well set apart, fine and flat bone, standing well up on toes and a good walker.

Skin: Free from coarseness, wrinkles or any spots.

Live weight: Mature boar generally weigh 270 to 360 kg.

Landrace

FIG.1.16 LANDRACE

The Landrace breed was developed in Denmark and is responsible for the Dane's pork producing name for which they are still famous. They produce the best quality bacon in the world. Danish government protected this breed and did not allow export until 1940 and then an agreement was signed allowing shipment of the surplus stock.

Colour: White in colour, although black skin spots, freckles are rather common

Ear: Heavy, forward and down, lop-ears

Carcass: Long, deep side, square ham, trim jowl. More lean than that of the meat breeds. Less back fat and lard.

Disqualification: Black hair, erect ears and less than six teats on a side.

The breed is noted for prolificacy and efficiency of feed utilization. In U.S.A, the breed has been extensively used for developing crossbred foundations. In India, Landrace stocks have been imported with the same objective. The American Landrace is longer than other breeds because of an extra vertebra.

Chester White

FIG. 1.17 CHESTER WHITE
Provided by National Swine Registry

This breed originated in Chester and also in surrounding countries of Delaware in Pennsylvania. The foundation stock included important pigs of English Yorkshire, Lincolnshire and Cheshire breeds.

Colour: The pigs are pure white. Spots on hair or any other colour disqualifies the animal. No breed of pig is more attractive when washed and presented for show ring than Chester White.

Size: The pigs are of large size and quiet in temperament.

Ear: Two-third of the ears are drooping.

Carcass: The carcass of this breed is not very meaty. They have small loin-eye area and too much back fat.

Disqualification: Swirls on the back and sides, or any colour other than white.

The sows are prolific, suckles well and make good mothers. Chester White is prone to sun burn.

Ohio Improved Chester (OIC)

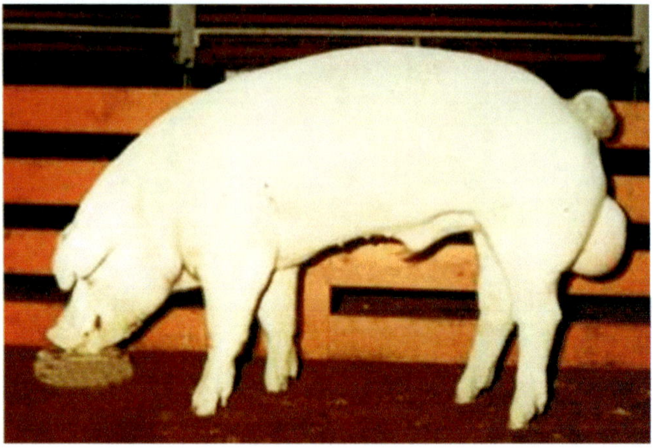

FIG. 1.18 OHIO IMPROVED CHESTER WHITE
Source: http://professorpaulanimalencyclopedia.blog.spot.com

The OIC breed was developed by L.B. Silver of Salem, Ohio in 1860. The size of the original Chester was improved to form this breed.

Black with some White

Hampshire

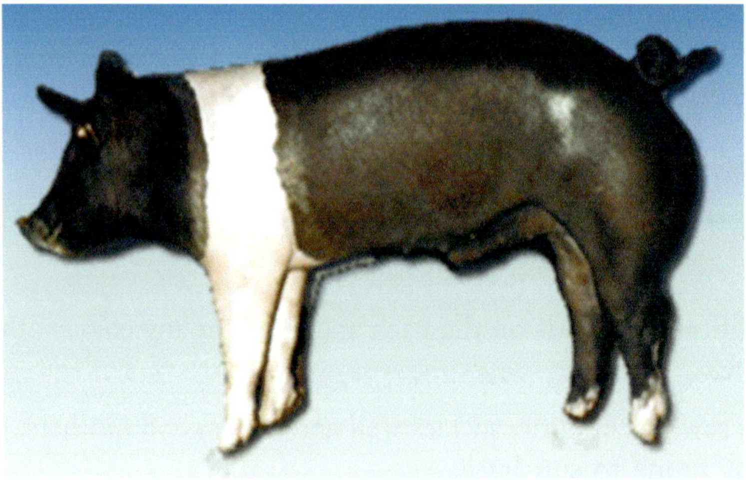

FIG. 1.19 HAMPSHIRE
Source: http://professorpaulanimalencyclopedia.blog.spot.com

The Hampshire breed originated in England. Major development of the breed occurred in Kentucky, where it was called 'Thin Rind' later in 1910, the name of the breed was changed to Hampshire. The sows are good mother, quite prolific and produce good litters.

Colour: The breed is black in colour with a white belt entirely encircling the body at the neck region, including both front legs and feet.

Size: The pigs are of medium size.

Ears: The ears are upright.

Carcass: Hampshire is known for excellent carcass quality with larger longissimus dorsi muscle and leaner carcass compared to Duroc and Yorkshire.

Disqualification: An animal that is solid black, white belt more than two-thirds back on the body, the belt does not completely encircling the body, white on the belly extending the full length of the body or one with swirls on the upper half of the body or neck is not considered true to the breed and may be disqualified.

The weak points of the breed are

1. The animal lacks length
2. The body weight gain is slow
3. Swirls are present
4. High incidence of cryptorchidism

Wessex Saddleback

Wessex Saddleback is essentially a bacon breed which originated from England. It is known for its prolificacy and has a robust make-up.

Colour: Head, neck, hind quarters, hind legs and tail of this breed are black.

Head: Head is fairly long with straight snout.

Ears: The ear is two-third drooping having a forward pitch.

FIG. 1.20 WESSEX SADDLE BACK
Source: http://professorpaulanimalencyclopedia.blog.spot.com

Berkshire

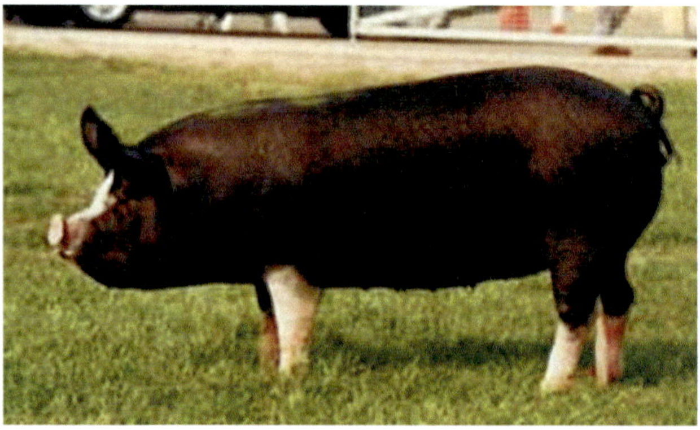

FIG. 1.21 BERKSHIRE
Source: http://professorpaulanimalencyclopedia.blog.spot.com

Berkshire is one of the oldest breeds of pigs tracing its origin in Berkshire and surrounding area of south-central England. The breed originated from crosses of Old English Hog, a descendant of wild boar, served as foundation stock and these animals were further improved by introducing Chinese, Siamese and Neopolitan blood. The breed is now valued as producer of meaty, well balanced carcass with a high cut-out value. In India, this breed was specially used for upgrading programme.

Colour: It is black with "six white points" that is white on feet, nose and tail. But more than 10% of white disqualifies it.

Head: It has a short head with dished face.

Ears: Its ears are erect.

Back: Long, level arched narrow back.

Legs: Straight and well set long legs.

Live weight: Full grown boar 275-375 kg; Sow 200-290 Kg.

Carcass: Its reputation throughout the world has been for meaty well balanced carcass with a high cut out value.

Disqualification: Swirls on the back or sides, large amount of white hair on the body, or red hair.

Poland China

FIG. 1.22 POLAND CHINA

Poland China is another breed with black colour and some white patch. The Poland China originated in the Miami valley of Ohio. No swine from Poland were used. The breed was designated as Poland China by a polish farmer Mr. Asher.

Colour: Poland China is black with six white points on the feet, face and tip of the tail.

Ear: The ear is two-third drooping.

Carcass: This is one of the largest American breeds. The breed was developed for and is known for heavy ham carcass and gainability. They produce carcasses with low back fat and large loin eyes. They are quiet by temperament and rough in constitution.

Disqualification: Less than six teats on a side, swirls on the upper half of the body, hernia, cryptorchids.

Black and White Breed

Spotted or spot

FIG. 1.23 SPOTTED
Source: http://professorpaulanimalencyclopedia.blog.spot.com

Spotted swine, previously known as Spotted Poland China, originated in Indiana, from the Poland China breed that is often spotted at birth. Later crosses were made with pigs from England called Gloucester Old Spot. Spots are spotted black and white with either black or white dominating. The body must not show less than 20% of either colour. Breeders prefer the spots black or white to be well-defined with no intermingling.

Size: Spots are large-sized pig, quiet in temperament.

Face: The face is dished.

Ears: Ears are two-third drooping

Hair coat: The hair coat is coarse

Disqualification: Brown or sandy spots, swirls on any part of the body, cryptorchids.

Sows are not very prolific.

Black Breeds

Large Black

FIG.1.24 LARGE BLACK
Source: http://professorpaulanimalencyclopedia.blog.spot.com

Large Black pig was developed in Britain. In the early part of this century, the Large Black were used for the production of pork in outdoor operations. Its hardiness and grazing ability make it an efficient meat producer. Large Blacks are also known for their mothering ability, milk capacity and prolificacy. Early selection was toward the bacon type and now the breed is characterised by great length and depth of the body.

Colour: The colour is completely black.

Head: long head and straight face.

Ears: The heavy looped ear is characteristic of the breed. In fact, the ears are so large that they cover much of the face and seem to obstruct the view from the eyes

Body: Large Blacks have a good depth of body, fair length, and strong backs.

The breed is known for its docile nature. The breed is said to have high levels of hardiness and thriftiness, with light shoulders, and good sides and hams. The mothering ability and litter sizes are good. The Large Black is listed as critically endangered by the American Livestock Breeds Conservancy.

Red Breeds

Tamworth

FIG.1.25 TAMWORTH
Provided by Michael von Luttwitz

The breed originated in Ireland. It derives the name from the town of Tamworth, which is located on the river Thames in Central England. It is one of the oldest of the purebred breeds.

Colour: Tamworth is red, with shades varying from light to dark.

Head: Head is strikingly long and narrow with a long snout.

Ears: They have fairly large ears that are carried somewhat erect.

Carcass: Long smooth sides and strong backs. Carcass produces bacon of the finest quality.

Legs: Individuals are long legged.

Live weight: Mature boar weighs about 350 kg and sow- 250-350 kg

The sows are prolific and careful mothers. They are excellent foragers. The breed has been widely used for cross-breeding in the tropics particularly in South East-Asia.

Disqualification: Swirls on the side and back and inverted teats.

Duroc

FIG 1.26 DUROC

A blending of two breeds (Jersey Red and Durocs of New York) in the North-Eastern United States gave rise to one of the most popular breeds the Duroc Jersey. The name Jersey was later dropped.

Colour: The breed is red in colour with shades varying from golden to cherry red, which is the most preferred colour.

Ears: The ears of Duroc are two-third drooping.

Carcass: The Duroc carcass is considered to be lean and excellent meat type.

Live Weight: Mature boars weigh 400 kg and sow 350 kg.

Disqualification: Swirls on the back and sides, black spots larger than two inches in diameter, cryptorchidism, less than six teats on a side or white hair on the body.

The Duroc is noted for excellent rate of weight gain and feed efficiency, maturing early. The Duroc sow has large litters and is a good mother. The breed is popular in South-East Asia and in American tropics mostly due to its colour, hardiness and fast growth.

Red with some White

Hereford

FIG. 1.27 HEREFORD
Source: http://professorpaulanimalencyclopedia.blog.spot.com

The Hereford is a breed whose colours are similar to Hereford cattle; white face, underline and switch. Red body with white on at least two feet and extends up to an inch or more above the hoof. This breed was developed in 1830 by R.U. Weber of Laplata, Missouri. In Size, the Hereford is smaller than other breeds of swine and in general is a compact type. Herefords are prolific, good mothers and have good foraging ability.

Colour: Red with white face.

Ears: The ears are forward drooping

Disqualification: No white on the face, less than two-thirds of the body red, swirls on the body, less than two white feet.

Swine Breeds Developed from In-bred Lines

New breeds of pigs were developed by cross-breeding. In all the cases two or more established breeds were crossed.

Minnesota No. 1

This breed was developed by winters *et al.* (1936) at the North Central Minnesota Agricultural Experiment Station. USDA.

Minnesota No.1 = 52% Tamworth x 48% Danish Landrace.

Representatives of this breed have a long body and a bit shorter legs than the average of most American breeds. The colour of this breed is red with a tinge of black and occasionally a few black spots. The ears fall somewhat forward. They are neither erect nor drooping. In many cases the tip of the ear reaches the tip of the snout which is fine, straight and medium length.

Minnesota No.2

Winters *et al.* (1941) at the Rosemount experimental station at Minnesota developed a new breed using Poland China and Canadian Yorkshire. The pigs of this breed are of somewhat shorter body but a bit longer than Minnesota No.1.

Minnesota No.2 = 40% Yorkshire x 60% Poland China.

The colour of the breed is black and white and the ears are erect. The pigs of this breed are temperamentally active.

Montana No.1

Montana No.1 = 45% Hampshire x 55% Landrace.

It was developed by the U.S. Department of Agriculture in the year 1936. Black Hampshire were used to fix the black colour.

Beltsville No.1

It was developed in 1935 at the USDA research station in Maryland.

Beltsville No.1 = 75% Danish Landrace x 25% Poland China.

The colour of the pigs is black with white markings.

Beltsville No.2

Three existing breeds of pigs were brought together to develop this breed. It was developed in 1952 at the USDA research station in Mary land. The colour of this breed is light red.

Beltsville No.2 = 58% Yorkshire x 32% Duroc x 5% Danish Landrace.

Maryland No.1

This breed was developed at the Queenstown Maryland Agricultural Experimental Station in 1941.

Maryland No. 1 = 68% Landrace x 32% American Berkshire.

The pigs are black and white spotted.

Palouse

This breed was developed at the Pullman, Washington Experimental Station in 1945. This breed was produced by crossing Danish Landrace and Chester White. The colour of the breed is white

San Pierre

This breed was developed by Gerald Johnson at Washington, Indiana in the year 1953 by crossing Berkshire and Chester White. The colour of this breed is black and white.

The exotic breeds of pigs with their characteristics are summarised in Table 1.1

TABLE 1.1 SUMMARIES OF SWINE BREEDS AND THEIR CHARACTERISTICS

Breed	Place of Origin	Colour	Distinctive Head Characteristics	Disqualifications
Berkshire	England, Chiefly in the south central countries of Berkshire and Wiltshire.	Black with 6 white points, 4 white feet, white on face and switch of the tail.	Medium short nose and erect ears.	A swirl on upper half of the body, fewer than 12 teats; cryptorchid or infantile vulva. Extreme pug nose is objectionable.
Chester White	U.S.A. chiefly in Chester and Delaware countries.	White. Small bluish spots are sometimes found on the skin, but are discriminated against.		Off-coloured hairs, cryptorchidism, hernia or swirls on body above flanks.
Duroc	U.S.A. chiefly in New York and New Jersey.	Red, Varying from light to dark.	Medium sized ear, tipping forward.	White feet or white spots on any part of the body, black spots larger than 2 inch in diameter, swirls on upper half of the body or neck, ridgling (one testicle boar), less than 6 teats on either side.
Hampshire	U.S.A. in Kentucky.	Black with a white belt around the shoulders and body, including the front legs.	Longer and straight face than most breeds; ears carried erect.	Too much white on ham, hind legs or head; or incomplete belt.
Hereford	U.S.A. by R.U Weber of La Plata Missouri	Red body colour, with white face, legs and switch similar to Hereford cattle	A white face and not less than two-thirds red, exclusive of face and ears	A white belt extending over shoulders, back or rump, fewer than two white feet, swirls on any part of the body, permanent deformities of any kind, less than 6 teats on each side.

[Table Contd.

Contd. Table]

Breed	**Place of Origin**	**Colour**	**Distinctive Head Characteristics**	**Disqualifications**
Landrace	Denmark	White although small black skin spots are common	Medium lop ears, straight snout, and trim jowl.	Black in the hair coat; fewer than 6 teats on either side; erect ears, with no forward bend.
Poland China	U.S.A.; Miami valley of Ohio.	Black with 6 white points- the feet, face and tip of the tail	Drooping ears	Less than 6 teats on a side, a swirl on upper half of body, hernia or cryptorchidism.
Spotted	U.S.A. ; chiefly in Indiana	Spotted black and white about 50% each		Brown or sandy spots; swirl on the upper one-third of the body; ridgeling; Solid black hair from ears forward; a combination of black leg and black hoof
Tamworth	England; in the central countries of Stafford, Leicester, Warwick and Northampton	Red varying from light to dark. Black spots may occur, but are objectionable	Wide between the ears, snout moderately long and straight, neat jowl, and medium sized erect ears.	Swirls, more than 5% black.
Yorkshire (Known as the Large White in England)	England	White, although black freckles may appear	Slightly dished face and erect ears	Swirls on upper third of the body, hair other than white, cryptorchidism, excessive amount of black or dark pigment on the skin. Less than 6 teats on each side.

Important Nomenclatures

Agalactia: Failure to secrete milk following parturition.

Average daily gain (ADG): Amount of weight gained per day per pig during a defined period of time.

Bacon: Pig meat sold after curing, which usually involves immersion in a solution of brine.

Barrow/ Hog: Male pig castrated before attaining sexual maturity.

Boar/ Barn/ Hogg: An un-castrated adult male pig kept for breeding purpose.

Boarling: Young male pig.

Cannibalism: The eating of young ones.

Card/Crit or Runt: The smallest pigling in a litter, usually the last to be farrowed.

Closed pig/ In pig: Young female pig kept for breeding, which is pregnant.

Coupling: The act of mating in pig.

Creep: An enclosure of feeder used for supplemental feeding of nursing young one, which is not accessible to the sow.

Cryptorchid/Ridgling: A male which is sterile because one or both testicles are retained in the abdominal cavity.

Dark, firm and dry (DFD): DFD muscle is attributed to poor pre-slaughter handling and depletion of muscle glycogen through excitement. The pH remains around or above 6.5. The inadequate levels of glycogen due to chronic stress at the time of slaughter leads to inadequate lactic acid therefore, failure to reach adequately low pH. This meat lacks flavour.

Drove/Stock or Herd: Group of pigs.

Farrowing: The act of parturition in pigs.

Farrowing crate: A steel crate used in the farrowing pen to contain the sow, so as to reduce overlaying piglet mortality at the same time allowing access to her teats.

Farrowing index: The number of farrowing per sow per year.

Farrowing interval: The average time for a sow or a group of sows between one farrowing and the next.

Farrowing rate: The number of sows farrowing out of the sows initially served, expressed as percentage.

Flushing: The practice of feeding female pigs generously two to three weeks before breeding.

Gilt/ Yelt: Young female pig kept for breeding purpose and up to the time of first litter. After farrowing she becomes Sow.

Grunting/ Squealing: The sound produced by pigs.

Guard rails/ pig fenders: Galvanised iron pipes or rails provided 25 cm above the ground level and 20 cm from the wall, all around the farrowing pen, to prevent loss of piglets due to crushing. (Piglets can run under these rails when the sow lies down so that they are not crushed between the sow and the wall)

Ham/ Gammon: The thigh of hog prepared for food. The hind leg of swine from the hock upward on the live animal.

Hernia: The protrusion of part of the intestine through an opening in the body wall.

Lean cuts: Ham, loin, boston butt and picnic.

Litter: The piglets farrowed by a sow in one delivery

Loin eye area (LEA): Loin eye area is measured between the tenth and eleventh ribs in swine. Only the large longissimus dorsi muscle is included in the measurement. The muscle is traced on a clear piece of acetate paper and the area is later determined using a plani-meter. It is the best indicator of the degree of muscling in the carcass. The loin eye area should be 4.5 square inch or more in meat-type hogs.

Mastitis Metritis Agalactia complex (MMA): It is a complex disease causing inflammation of mammary glands, inflammation of uterus and complete lactational failure. The MMA syndrome can involve, metabolic, bacterial and hormonal factors with stress playing a part. The symptoms are anorexia, lethargy, disinterest towards piglets, high fever, swelling of udder, hard to touch, agalactia and mucopurulent vaginal discharge.

Meconium: Excrement accumulated in the bowels during foetal development.

Mummified foetus: A shrivelled or dried foetus that has remained in the uterus instead of being expelled or aborted after dying.

Needle teeth: Eight small, sharp tusks like teeth, which pigs have at birth.

Open gilt: Young female pig kept for breeding, which has not yet been served.

Pale Soft Exudates (PSE): It is a condition of the pork immediately after slaughter as a result of rapid postmortem pH decline while the muscle temperature is high. The combination of low pH and high temperature causes both myofibrilar and sarcoplasmic protein to denature. This denaturation results in a paler colour, a soft open structure as well as a poor water holding capacity, which results in drip (exudates) as high as 10-15%.

Piglets/ pigling: Newly born un weaned pig of either sex, usually less than eight weeks of age

Porcine Stress Syndrome (PSS) or malignant hyperthermia: PSS is associated with the halothane sensitive gene and with rapidly growing meaty pigs. Pigs will show an acute temperature rise, muscular rigidity, sudden death and a blueing of the extremities. Muscle will be pale, soft and exudative. These symptoms usually occur after exposure to acute stressful conditions.

Pork: Pig meat sold in a fresh state.

Primal cuts: Ham, loin, boston butt, picnic and bacon.

Service period: The interval between weaning of the sow and re-mating.

Shoat: Young swine of either sex, generally weighing less than 50 Kg.

SMEDI: It is a complex disease in swine causing still birth, mummification, embryonic death and infertility. Now SMEDI is identified as a viral agent causing all these conditions in pigs. Control measures are by good management practices.

Soft pork: In swine, the quality of body fat changes with the dietary fats provided. Feeds like groundnut, soybeans, rice bran, vegetable oils etc., when fed in liberal amounts in finishing stage results in deposition of large amount of short chain fatty acids similar to those present in these feeds. Soft pork is slimy to touch and unattractive. The meat is difficult to handle or cut since the oil drips from the meat. There will be more shrinkage when it is smoked and more wastage during storage and cooking.

Sow: Adult breeding female pig after first litter.

Stag/ Steg: Male pig castrated later in life. (After reaching sexual maturity)

Still born: Born dead at birth.

Store pig: Pig between weaning and fattening generally from 8 to about 12 weeks of age.

Sus/Swine: Species of pigs.

Weaning: Removal of young ones from their dam so as to prevent them sucking their mother.

Functions of Swine

The primary purpose of pig farming all over the world is the production of meat, including pork, bacon or fat. Secondary considerations are the production of pig skin, bristles and manure. The important functions of swine are:

1. Contributes food

They produce more live weight gain from a given weight of feed than any other class of meat animals. The protein of pork provides all of the essential amino acids including, lysine and methionine, along with minerals and vitamins. 97% of the pork protein and 96% of meat fats are digested by human beings. Pork supplies $^{2}/_{3}{}^{rd}$ of the requirement for vitamin B-12, a vitamin which occurs only in animal source and fermentation products. Iron in meat is about twice as high as in plants. But pork is deficient in calcium, vitamin A and vitamin D.

TABLE 1.2 CHEMICAL COMPOSITION OF PORK

Moisture	74.3 %
Protein	18.6 %
Fat (ether extract)	6.0 %
Minerals	1.1 %
Calcium	0.03 %
Phosphorus	0.20 %
Iron (mg)	2.3 %
Calorific Value (per 100g)	134
Vitamin A (I.U per 100g)	0
Vitamin B (I.U per 100g)	60

Source: Zullo *et al.*, 2006

Pork is higher in polyunsaturated fatty acids than beef, lamb, or milk and is lower in cholesterol than butter, cheese, eggs and many sea foods.

2. Provides profitable returns from available labour

Swine enterprises provides for relatively high returns from labour in comparison with other types of livestock production and the labour requirement is uniform throughout the year.

3. Convert inedible feeds into valuable products

Swine are better adapted than any other class of livestock for utilising several wastes and agricultural by-products that are not suitable for human consumption and convert it into high quality animal protein suitable for human consumption.

4. Aids in maintaining soil fertility

Pig manure is useful as fertilizer for soil or for fish ponds; for production of methane gas; and for the culture of algae such as *chlorella* and *azolla* that can be used as animal feed. The pig manure contains, 0.7% Nitrogen, 0.68% Phosphorus and 0.7 % Potassium.

5. Serves as an important companion of grain production

By-products of grain production like bran; polish *etc.* can be converted to protein by pigs.

Thus swine gives elasticity and stability to grain production.

6. Supplements other enterprises

Skins: Pig skin can be used for shoe manufacture, belt and other leather goods.

The fats: Inedible fats are used in the manufacture of soaps, animal feeds, binders, lubricants, fertilizers, shaving cream *etc.*

Variety meats: The liver, brain, kidney, stomach, ear, skin, snout and weasand (oesophagus) are used as variety meats. They represent about 3.25% of the live weight of pigs.

Hair: Pig bristles are used for making brushes, mattress, and insulation boards.

Hoofs: Hoofs are converted into fertilizer and plaster retarder.

Blood: Used in the refining of calico, and for making shoe polish.

Meat scrapes: Meat scrapes are made into meat meal or tankage.

Bones: Bones are converted into fertilizer, glue, knife handles *etc.*

Intestine and bladder: Used for preparation of lard and for producing putty and casings.

Endocrine glands: Endocrine glands are used in the manufacture of numerous pharmaceutical preparations – Thyroid, parathyroid, pituitary, pineal, adrenal, pancreas *etc.* can be used.

Collagen: Made into glue and gelatine. Gelatin is used in ice-creams, capsules, photography, culture media and for making explosives.

7. Pigs in transgenics and for xenotransplantation

Swine are increasingly used as models in bio-medical research. Developments in the field of immunology and xenotransplantation prospects have renewed interest in this species, the anatomy and physiology of which have notable similarities to those of man. The first transplantation of hog heart valve was carried out in 1971. Till now 3,50,000 hog heart valves have been implanted in humans.

Advantages of Pig Farming

1. Pigs are efficient converters of feed into pork and pork products which have a high nutritional value in human diet. The feed conversion ratio is more efficient in pigs compared to other livestock (3 kg feed into 1 kg meat compared to 5 kg feed into 1 kg meat in beef).
2. Pigs are prolific breeders, grow rapidly, mature quickly and can be managed to produce more than two litters in a year with 8 to 12 piglets in a litter.
3. Gilts can be bred at the age of 8-9 months.
4. Gestation period is less (114 days or 3months, 3 weeks, 3 days)
5. Pigs can attain a live body weight of 90 kg in 6-8 months.
6. Large acreages of unutilised land in the country can be utilised for raising pigs.

7. The pig enterprise is flexible and farmer can increase the size of his farm as profit increases.
8. The pig farming requires moderate investment than other farming.
9. It creates profitable work for the available labour on the farm as it can be successfully combined with dairy and other agricultural activity.
10. Pig utilise the waste products very efficiently. Table garbage, bakery waste, hotel and kitchen waste and unmarketable fruits and vegetables can be fed to pigs.
11. Pigs aid in maintaining soil fertility by the production of manure of good biological value.

Constraints of Pig Farming

1. Pig rearing was confined with the traditional pig keepers who where socio-economically very poor belonging to weaker sections of the society.
2. Social taboo for pork consumption due to religious inhibitions.
3. Negligence of indigenous pigs and letting them loose for subsistence through scavenging.
4. Poor and primitive system of rearing pigs in unhygienic conditions without any proper sanitary measures.
5. Non-existence of any progressive program to improve indigenous stock.
6. Absence of marketing facility and low consumer demand for pork products.
7. Constraints like inadequate number of high quality stock for distribution among pig breeders.
8. Lack of balanced feed at economic price to the common pig breeders.
9. In-adequacy of relevant data and scanty information on economics of pig farming.
10. High feeding cost accounting for about 70-80% of total cost of pig production.
11. Poor performance during rearing and growing period leading to low production below their production potential.

12. Lack of proper housing facilities.
13. High pre-weaning mortality and morbidity.
14. Pigs are susceptible to numerous diseases and parasites.

Scope of Swine Farming in India

The increased pressure of the growing Indian population and the rising living standards has created the scope for the production of more animal products. The meat production from buffalo, sheep and goats are unlikely to keep pace with the rising population because of low prolificacy and high generation interval. The expected demand for meat and meat products can be met mainly by increased pig and pork production. Eggs and pig meat are the important source of nutrients in the human diet because of their better nutritional value and utilisation. Pig meat is an important source of animal protein, vitamins minerals and other micro-nutrients. Indian pig industry can play a major role in the supply of high quality nutritious diet. Pig production is important for the future development of Indian meat industry as pig is the most efficient feed converter among domesticated livestock. Pig rearing fits in well with mixed farming and can also be complementary to intensive crop production operations. In India, pig farming has a special significance and can play an important role in improving the socio-economic status of sizable section of the weaker community as pig rearing is very popular amongst tribal populations. Government has extended loans for pig farming through various schemes to uplift weaker sections of the communities through Special Livestock Breeding Programme (SLBP), Intensive Rural Development Programme (IRDP) and Prime Minister's Rosgar Yojana (PMRY) *etc.*

Women can play a significant role in the rural pig production for their socio-economic up-liftment. Pigs provide pin money to the house wives and help her to look after the family's nutritional need. Pig rearing offers opportunity for women's employment. Pig farming needs low investment, simple skills and less land. It can also be practiced as a cottage industry.

Feed accounts to about 80% of the total cost of pig production. Efficient use of cheaper rations based on locally available materials and agricultural by-products are essential for profitable pig production. Modern technology of pig feeding consists of meeting the biological needs of the pigs with proper combination of feed ingredients and locally available by-products to yield economic production of pigs and piglets.

India has excellent possibilities for a flourishing pig industry. Stimulation of this industry can solve the problem of poverty and malnutrition to a great extent. It improves utilisation of agricultural by-products, diversify agricultural production and employ available man power resources.

Demography of Pig Population

As per FAO stat-2018, the world pig population in the year 2016 was 981.79 million which represents an increase of 324 million since 1966-67.

- China alone has 456 million pigs, which accounts for 83% of Asian pigs.
- In Europe, Germany contributes the largest number having 27.37 million pigs.
- In North Central America, the USA has the largest number of pigs 71.5 million.
- Mexico has 16.75 million pigs.
- Canada has 12.77 million pigs.
- In South America, Brazil alone have 39.95 million pigs.
- In developing countries, pig population increased by 21%.
- In developed nations pig population decreased by 3%.
- Developing countries have more than 62% of total world pig population.

The trends of world pig population is depicted in Fig. 1.28 and global distribution of pigs from 2009-2016 in Table 1.3

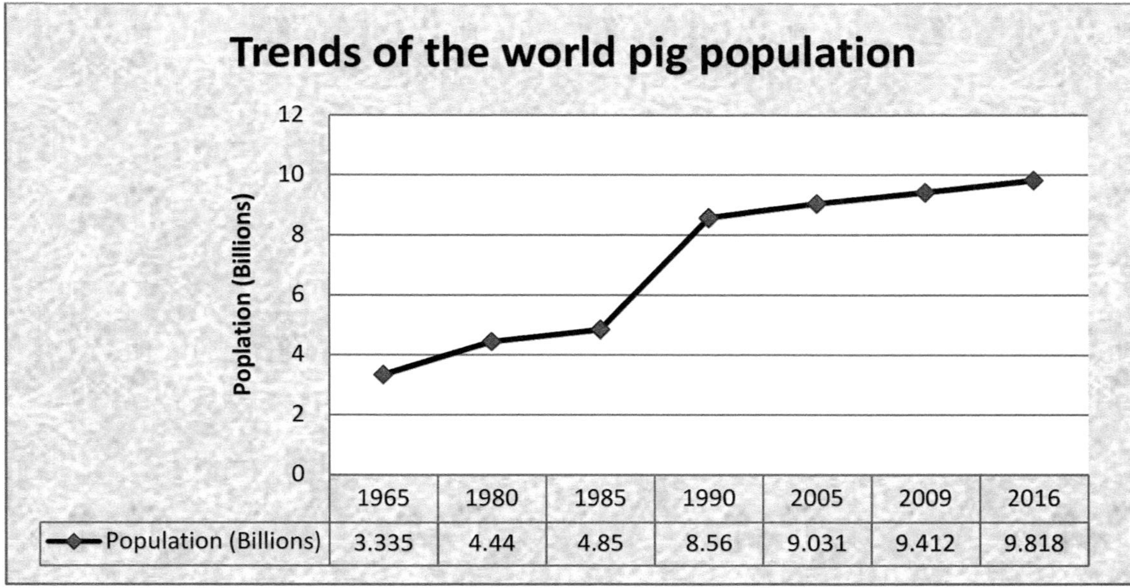

FIG 1.28. TRENDS OF THE WORLD PIG POPULATION

TABLE 1.3 GLOBAL DISTRIBUTIONS OF PIGS 2009-2016

COUNTRY	PIG NUMBER IN MILLION		
YEAR	**2009**	**2016**	**Per cent of the world**
AFRICA	27.49	36.62	2.2%
NORTH & CENTRAL AMERICA	199.28	179.09	10.7 %
U.S.A	67.14	71.5	6.7%
CANADA	12.18	12.77	1.2%
MEXICO	15.20	16.75	1.9%
SOUTH AMERICA	6.2	6.91	6.2%
BRAZIL	38.04	39.95	3.8%
ASIA	561.22	573.64	52.4%
CHINA	456.8	456.67	43.9%
VIETNAM	27.62	29.07	1.4%
JAPAN	9.89	9.31	1.3%
EUROPE	187.74	186.99	20%
GERMANY	26.88	27.37	3.0%
POLAND	14.27	10.86	2.6%
SPAIN	26.28	29.23	2.0%
NETHERLANDS	12.10	12.47	1.6%
FRANCE	14.81	12.70	1.4%
DENMARK	12.3	12.38	10.3%
OCEANIA	5.2	5.43	4.1%
RUSSIAN FEDERATION	16.16	21.5	4.1%
UKRAINE	6.52	7.07	2.1%
TOTAL	941.2	981.79	

Source: FAO Stat-2018

Pork is currently the most consumed meat in the world, at present (15.9 kg per person), when compared to that of poultry (14.6 kg per person) and bovines (9.4 kg per person). World pork production in 2016 was 118.168 million tonnes, of which 56% was within Asia. Europe comes second with 25% and America with 17%. The four largest world producers are China with 52.75 million tonnes, The European Union (EU 25) with 23.45 million tonnes, the U.S.A. close to 11.84 million tonnes and Brazil, with 3.82 million tonnes. These four largest world producers of pork constitute 80% of the world production.

Top ten countries in pig number and ten largest world's pork producers are shown in Table 1.4 and 1.5.

TABLE 1.4 TOP TEN COUNTRIES IN PIG NUMBER – 2017

RANK	COUNTRIES	PIG NUMBER (Millions)
1	CHINA	435.04
2	EUROPEAN UNION	147.24
3	U.S.A	68.92
4	BRAZIL	39.23
5	SPAIN	29.23
6	VIETNAM	29.07
7	GERMANY	27.37
8	RUSSIAN FEDERATION	21.5
9	CANADA	12.77
10	FRANCE	12.70

Source: FAO Stat-2018

TABLE 1.5 TEN LARGEST WORLD PRODUCERS OF PORK, 2017

RANK	COUNTRIES	PORK PRODUCTION (Million Tons)
1	CHINA	52.750
2	EU-25	23.450
3	U.S.A	11.844
4	BRAZIL	3.815
5	RUSSIA	3.000
6	VIETNAM	2.750
7	CANADA	1.950
8	PHILIPINES	1.610
9	MEXICO	1.420
10	SOUTH KOREA	1.305

Source: USDA Foreign Agricultural Service 2018

Export and Import

The largest exporters of pig meat globally are the U.S.A, Denmark and Germany, accounting for over 50% of the gross exports. The largest exporters of live pigs globally are the Netherlands, Denmark and Canada, accounting for over two thirds of gross exports. The largest importers of pig meat globally are Germany, Mexico and the U.S.A, accounting for nearly a third of gross imports. The largest importers of live pigs globally are Germany and the U.S.A accounting for over half of gross imports between them (Table 1.6).

TABLE 1.6 TOP TEN EXPORTING COUNTRIES OF PORK, 2017

RANK	COUNTRIES	PORK PRODUCTION (Million Tons)
1	U.S.A	12.40
2	DENMARK	11.57
3	GERMANY	8.53
4	BELGIUM	4.59
5	SPAIN	4.04
6	NETHERLAND	3.37
7	FRANCE	3.06
8	POLAND	1.75
9	CANADA	1.55
10	BRAZIL	0.69

Source: USDA Foreign Agricultural Service 2018

Per capita meat consumption is expected to increase from the present 40 kg to 45 kg in 2030 (Roppa, 2006). The greater part of the increase in world demand for meats in the next decade will occur in the developing countries. In general, animal production will occur more in developing countries with favourable climate, land, human resources and in countries which will produce food with quality, safety and low cost. Factors associated with public attitudes, government policies, environmental issues, global or local trading conditions, food versus fuel scenario, or even climate change may provide the limitations to pig health and production in the future.

Indian Swine Scenario

India with a pig population of 10.29 million in the year 2012 formed only 1.05% of the world pig population of 981.79 million. Majority of these pigs (nearly 78.5%) are indigenous and the remaining 21.4% are crossbreds. The population of rural and urban pig is 91 and 9% respectively. The pig population in India has increased from 4.40 million in 1951 to 13.52 million in 2003 and then decreased to 11.13 million in 2007 and then to 10.29 in 2012 (Fig. 1.29). Pigs had the highest annual compound growth rate of 4.10% among livestock. Trends of Indian pig population is depicted in Fig. 1.29 and growth pattern of pigs in India is given in Table 1.7, followed by state wise population trends in Table 1.8.

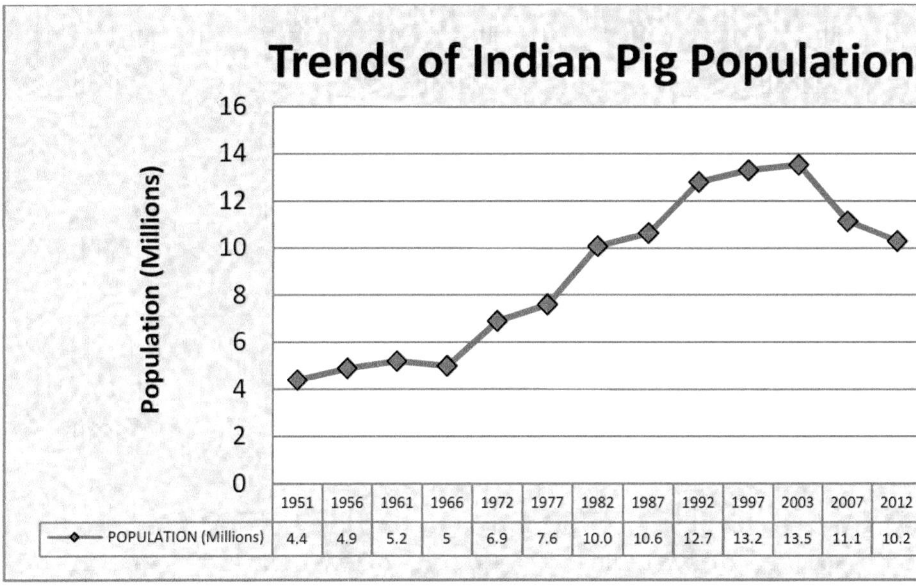

FIG. 1.29 TRENDS OF INDIAN PIG POPULATION

TABLE 1.7 GROWTH PATTERNS OF PIGS IN INDIA 1951-2007

Year	Annual Compound Growth Rates (%)
1951-56	11.36
1956-61	6.12
1961-66	-3.85
1966-72	6.65
1972-77	38.0
1977-82	32.50
1982-87	5.56
1987-92	20.32
1992-97	3.91
1997-2003	1.73
2003-2007	-17.66
2007-2012	-7.54

Source: Basic Animal Husbandry and Fisheries Statistice-2017

TABLE 1.8 STATE WISE POPULATION TRENDS IN INDIA 2003-2012

	STATE	PIG NUMBER IN THOUSANDS			
		2003	2007	2012	Annual Growth Rate (%)
1.	Andhra Pradesh	570	439	394.36	-10.16
2.	Arunachal Pradesh	330	356	356.35	0.09
3.	Assam	1543	2000	1636.02	-18.20
4.	Bihar	672	632	649.79	2.81
5.	Chhattisgarh	552	413	439.06	6.30
6.	Goa	87	58	43.57	-24.87
7.	Gujarat	351	22	4.28	-80.54
8.	Haryana	120	134	126.95	-5.26
9.	Himachal Pradesh	3	2	5.03	151.5
10.	Jammu & Kashmir	2	1	2.42	142
11.	Jharkhand	1108	732	962.37	31.47
12.	Karnataka	312	281	304.8	8.46
13.	Kerala	76	59	55.78	-5.45
14.	Madhya Pradesh	358	193	175.25	-9.19
15.	Maharashtra	439	327	325.76	-0.37
16.	Manipur	415	314	277.22	-11.71
17.	Meghalaya	419	524	543.38	3.69
18.	Mizoram	218	267	245.24	-8.14
19.	Nagaland	644	698	503.69	-27.83
20.	Orissa	662	612	280.32	-54.19
21.	Punjab	29	26	32.22	23.92
22.	Rajasthan	338	209	237.67	13.71
23.	Sikkim	38	35	29.91	-14.54
24.	Tamil Nadu	321	284	183.98	-35.21
25.	Tripura	209	264	362.53	37.32
26.	Uttar Pradesh	2284	1350	1334.39	-1.15
27.	Uttarakhand	33	20	19.91	-0.45
28.	West Bengal	1301	815	648.11	-20.47
29.	Andaman & Nicobar	52	48	35.92	-25.16
30.	Chandigarh	0.29	0.27	0.14	-48.14
31.	Dadra & Nagar Haveli	3	0.18	0.00	-100
32.	Daman & Diu	0.17	0.01	-94.11	–
33.	Delhi	28	20	76.35	281.75
34.	Lakshadweep	–	–	0.00	–
	All India	13519.3	11134	10293.70	-7.54

Source: Basic Animal Husbandry and Fisheries Statistics-2017

Ten Indian states had more than 0.5 million pig. Maximum growth rate of pig population during the period 2003-2012 occurred in Assam followed by Uttarakhand. Maximum reduction of pig population during the same period occurred in Gujarat. Eastern region of the country comprising of Assam, Manipur, Meghalaya, Mizoram, Nagaland, Bihar, Orissa, Sikkim, Tripura and West Bengal had largest pig population (4.16 million), followed by Southern and Northern regions with 3.15 and 1.83 million pigs respectively. The population density of pigs was maximum in Nagaland (21.5 per Sq. Km) followed by Meghalaya and Manipur. Pig population per 1000 person in Eastern, Southern, Northern and Western regions of the country comes to 20, 14.2, 12.3 and 3.7 respectively.

According to the Livestock Census, 2012, published by Government of India (GOI), the pig population declined by 7.5 per cent to 10.3 million from 2007 to 2012. The decline in population may be attributed to disease outbreaks. The Eastern and North-Eastern regions of the country comprise around 63 per cent of the pig population. The highest pig population is in the state of Assam (1.63 million) followed by Uttar Pradesh (1.33 million), Jharkhand (0.96 million), Bihar (0.65 million) and West Bengal (0.65 million). The majority of the pig population in India is of indigenous breeds (76 per cent) though population of cross-bred and exotic pigs increased by 12.7 per cent from the year 2003 to 2012. The exotic breeds mainly comprises of Hampshire, Large White Yorkshire, Duroc, Landrace, and Tamworth while some of the popular indigenous pig breeds include Ghungroo, Niang Megha, Ankamali, Agonda Goan, and Tany Vo.

Pork production in India is estimated at 464 thousand metric tons in year 2014-15 (April-March) (GOI), which contributes approximately 8% of the country's animal protein sources. From the year 2009-10 to 2014-15, pork production increased at a slow pace with compounded annual growth rate of 1.4 per cent due to population growth. The pork production is concentrated mainly in the states of Uttar Pradesh

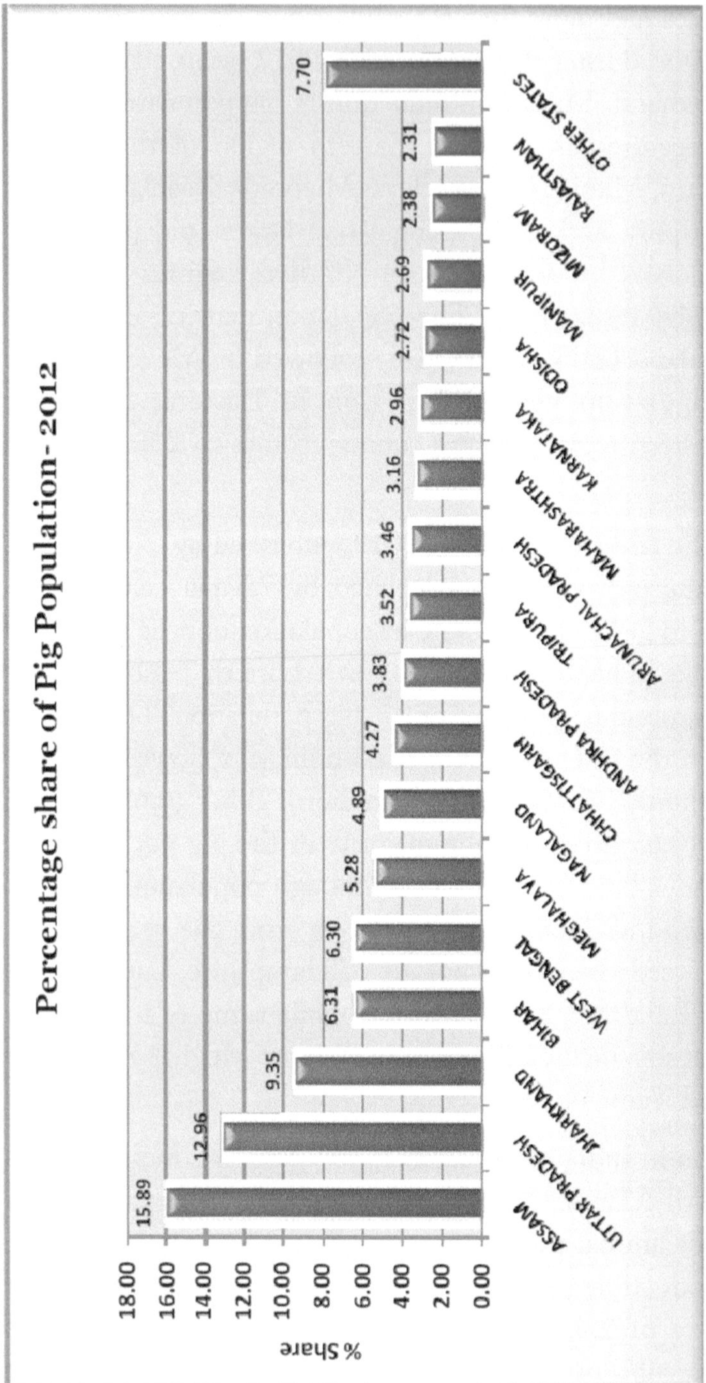

FIG1.30 STATE WISE PIG POPULATION-2012

Source: 19th Livestock census-2012, All India Report.

(30 per cent), followed by North-Eastern states (25 per cent), Bihar (15 per cent), West Bengal (6 per cent), Karnataka (4 per cent), Jharkhand (4 per cent), and Kerala (3 per cent). The state wise percentage share of pig population as per 19th livestock census-2012 is depicted in Fig. 1.30 and major production states in India are depicted in Fig. 1.31. The per capita pork consumption in India is negligible with the consumption mainly concentrated in North-Eastern states including Assam, Nagaland, Arunachal Pradesh, Manipur, Meghalaya, Mizoram, Sikkim, and Tripura. Other Indian states with high pork consumption include Bihar, Jharkhand, West Bengal, Goa and Kerala.

In fact, India's pork consumption can be divided into two segments. First segment being the consumption in the form of fresh pork meat sold through unorganised wet markets and meat vendors. The second segment is the high value imported pork products like sausages, ham, bacon, salami, canned meat products and frozen meat. The hotels and restaurants are the major buyers of the imported pork products, which cater to international travelers and wealthier Indian consumers.

The processed pork segment is still very small but growing at a rapid pace due to socio-economic and demographic changes. The major market for processed pork is limited to large Indian cities such as Delhi, Mumbai, Bangalore, Kolkata, and Pune. A large segment of the processed pork demand is in high-end hotels and restaurants which serve pork products such as ham, bacon, and sausages for the breakfast buffet and other meals.

Pork constitutes 6.5% of the total meat production of India. About 20% pigs are annually slaughtered for meat in India, which accounts for 6% of the total meat produced in the country. The highest per capita pork consumption was in Nagaland. The consumption of pork has increased in the recent years due to nutritional awareness among people.

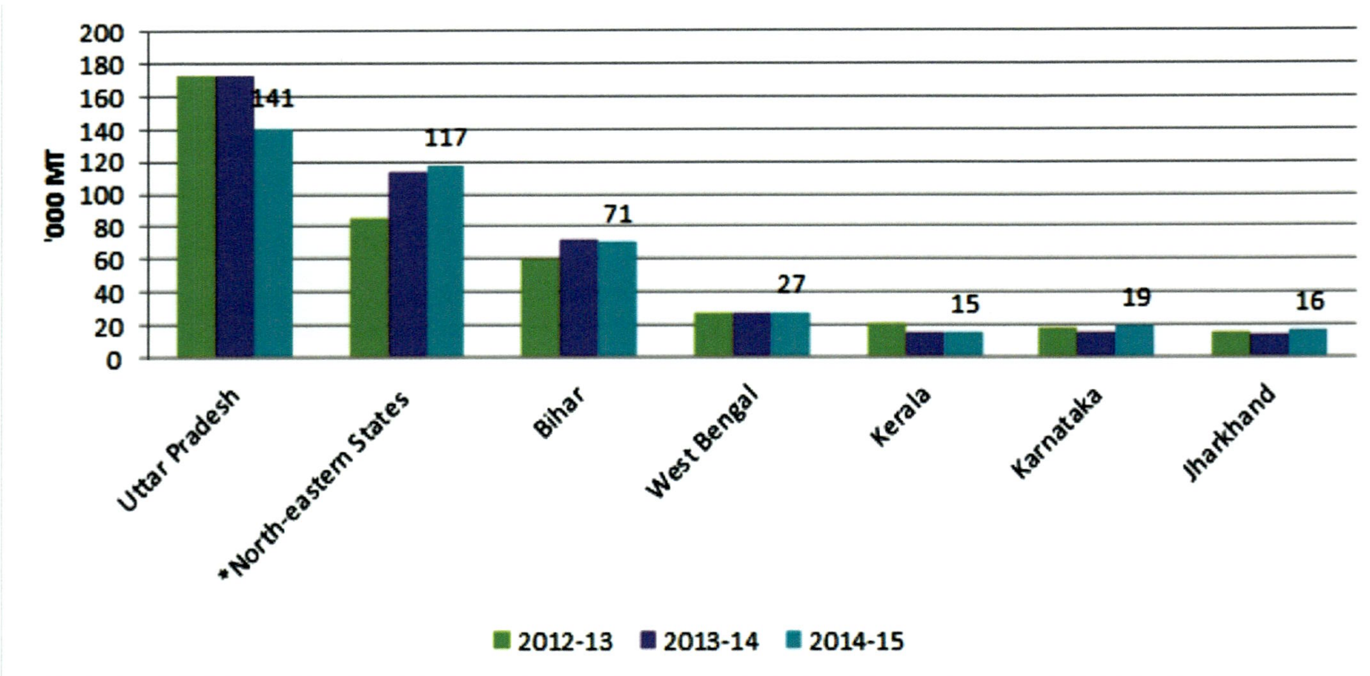

FIG 1.31 MAJOR PORK PRODUCTION STATES IN INDIA
(* North-eastern states include smaller hilly states of Arunachal Pradesh, Assam, Manipur, Meghalaya, Mizoram, Nagaland, Sikkim, and Tripura).
Source: Global agriculture information network report number. IN6098

TABLE 1.9 PRODUCTION OF PORK IN INDIA 1982-2016

Year	(Thousand Tonnes)
1982	80
1987	80
1992	397
1997	462
2003	462
2008	481.11
2015	387.55
2016	468.80

Source: Basic Animal Husbandry and Fisheries Statistics-2017

The pork consumption is predicted to be 499 thousand tonnes per year in India at an average consumption of 0.46 kg/person/year (Roppa, 2006). Piggery enterprise contributes about 8.9% of the total revenue from livestock production. According to the Basic Animal Husbandry and Fisheries Statistics-2017, published by the Department of Animal Husbandry and Dairying, Government of India, the total pork output was 468.80 thousand tonnes during the year 2015-16 (Table 1.9). The contribution of pork to GDP at current prices was Rs 5133 crores during 2015-16. 504.93 tonnes of pork meat was exported during this period which fetched a revenue of Rs. 275.88 Lakhs.

In 2015, pork products imports increased by 28% from the previous year to 527 metric tonnes. Major suppliers of pork meat products to India were Belgium, Sri Laka, Spain, Italy and Netherlands. Indian export of pork and pork products are negligible.

Challenges of Indian piggery sector

The major challenges that affect the growth of pork sector include: lack of sufficient breeder farms, deficiency of feed and fodder resources, diseases like classical Swine Fever, Porcine Reproductive and Respiratory Syndrome (PRRS), and Porcine Rotavirus, limited availability of vaccines, and insufficient slaughter and processing facilities across the country. The sector is also constrained as most of the pig farmers belong to the lower socio-economic strata and undertake pig farming as a livelihood rather than scientific pig farming with improved foundation stock, proper housing, feeding and management. Generally, pigs are fed concentrated feed with other locally available agro by-products, tuber crops like sweet potato, tapioca, colocasia, vegetables and kitchen waste.

Piggery Development Programmes in India

Piggery development has played a major role in solving the food problems of various thickly populated countries including China. In India, except for a few imports of superior quality exotic breeds of pigs by some missionary organisation, no organised measures were taken to improve pig production before the implementation of the five year plans. Pig development did not attract much attention in the first five year plan.

Second Five Year Plan (1956-61)

In order to develop pig production on systematic lines and organise it into a profitable industry to the farmers, a co-ordinated Piggery Development Programme was launched in 1959-60. Under this programme, by the end of second five year plan, 2 – regional pig breeding stations, 2 – bacon factories, 15 – pig breeding units and 33 – Piggery development blocks were established.

Third Five Year Plan (1961-66)

During the third five year plan, 6 - bacon factories (Table 1.10), 5 – regional pig breeding stations, 37 – pig breeding farms and 107 piggery development blocks were established. The primary objective of the regional pig breeding station was to acclimatise and further improve the breeding stock of exotic pigs through selective breeding and for distribution to pig breeding farms/unit. The pigs produced at the farms /units were distributed in rural areas for grading up the local stock. The bacon factories were to provide ready market for the pigs raised.

The exotic breeds introduced were Large White Yorkshire, Tamworth, Hampshire and Landrace.

TABLE 1.10 LISTS OF BACON FACTORIES IN INDIA

Sl. No	State	Capacity No. of pigs/day	Address
1.	Uttar Pradesh	100	Bacon Factory, Central Dairy Farm, Aligarh
2.	West Bengal	20	Bacon Factory, Haringhatta-721436, Mohanpur, Nadia
3.	Andhra Pradesh	100	Government Bacon Factory, Gannavaram, 521101, Krishna District
4.	Jharkhand	50	Government Bacon Factory, Kanqe- 834005, Ranchi
5.	Maharashtra	100	MAFCO Bacon Factory, National Park Borivilli, Mumbai- 400022
6.	Rajasthan	50	Meat Complex, Alwar – 301001
7.	Kerala	50	Meat Products of India, Koothattukulam-686662
8.	Punjab	20	Pork Processing Plant, Kharar – 140301.

Source: Annual Report of A.I.C.R.P on pigs 2001

Fourth Five Year Plan (1969-74)

Phase-I: Systematic research works on pigs was initiated by I.C.A.R with the launch of All India Co-ordinated Research Project (A.I.C.R.P on Pigs). The main objective of the project was to study the performance of existing exotic breeds under different agro-climatic conditions. The project started functioning from 1971 at four centres: Indian Veterinary Research Institute (I.V.R.I), Izatnagar as co-ordinating unit, Jawaharlal Nehru Krishi Vishwa Vidyalaya (J.N.K.V.V), Jabalpur, Acharya N.G Ranga Agricultural University, Tirupathi and Assam Agricultural University, Khanappara involving Landrace and Large White Yorkshire.

Phase-II: During the second phase, ICAR decided to exploit native germplasm of Desi pigs and there by remodelling the pig development programme with emphasis on inclusion of indigenous pig stock.

Subsequently emphasis was laid on crossbreeding of indigenous gilts to the boars of exotic breeds adapted in the various agro-ecological zones.

Fifth Five Year Plan (1974-79)

A.I.C.R.P was continued with the objective of evaluation of the performance of different breeds and breed combinations under different agro-climatic environment of the country. Exotic breeds like Large White Yorkshire, Middle White Yorkshire, Landrace, Tamworth and Wessex Saddle back and Hampshire where maintained at government farms with the objective of further improvement of the stock and distribution to farmers.

Sixth Five Year Plan (1980-85)

Exotic pigs were maintained at 99 state government pig breeding farms in the country. Better quality piglets were produced and distributed to the farmers for rearing and for establishment of small breeding units.

Seventh Five Year Plan (1985-90)

A.I.C.R.P on pigs was continued during 7th plan also. State government farms continued to maintain exotic breeds of pigs for distribution to farmers.

Annual Plan (1991-92)

A centrally sponsored scheme "Assistance to states for Integrated Piggery Development" by Ministry of Agriculture, Department of Animal Husbandry was started during 1991-92, with the objective to establish new pig breeding farms and to strengthen existing pig breeding farms with a view to produce more number of exotic pigs for distribution to the farmers.

Eighth Five Year Plan (1992-97)

During the 8th plan period, Rs. 851 Lakh was released for piggery development under the scheme "Assistance to states for Integrated Piggery Development". The A.I.C.R.P on pigs was continued with broadened objectives of developing improved breeds of pigs for different localities and to evolve economic rations. During 1993, two more centres were added: Kerala Agricultural University, Mannuthy and Tamil Nadu Veterinary and Animal Sciences University, Kattupakkam. Other schemes that were implemented during that period where:

a) **Grading-up Scheme**: It consisted of provision of an exotic boar for grading up of local pigs with a subsidy given for feeding and maintaining the boars.

b) **Fatteners Scheme**: Distribution of weaned piglets at subsidised rate at nominal charges, which will be purchased back by the bacon factories at a fixed rate.

c) **Intensive Rural Development Programme (I.R.D.P)** The scheme ran with collaboration of nationalised banks covering the entire state and provided subsidy for pig rearing for all classes of people. Through this scheme, three to four piglets of both sexes were distributed to pig breeders below poverty line.

d) **Special Livestock Breeding Scheme (S.L.B.P)**: This scheme was also implemented with the collaboration of nationalised banks in which pig breeders were provided with a unit of either three or four piglets.

e) **Special Animal Husbandry Programme:** The scheme was meant for piggery development block. The farmers were given 50% subsidy by the state and the rest 50% loan from nationalised banks for rearing pigs.

Ninth Five Year Plan (1997-2002)

Under the scheme of "Assistance to states for Integrated Piggery Development", an amount of Rs. 400 lakh was released to strengthen

34 pig breeding farms during 1997-98, while Rs. 400 lakh was released to 38 pig breeding farms during 1998-99. Assistance was also given to 13 new pig breeding farms at 10 K.V.K. The A.I.C.R.P on pigs has made significant contribution in research on breeding, nutrition and management aspect. During the ninth plan two more centres, Birsa Agricultural University, Ranchi (Jharkhand) and I.C.A.R research complex, Goa had been added. The total outlay of this project during the ninth plan was Rs. 874.10 lakh. I.C.A.R opened the National Research Centre on Pigs (N.R.C Pigs) at Rani, Guwahati. Besides this, ICAR has launched the National Agricultural Technology Programme (N.A.T.P) Mission Mode on Pigs with 100% I.C.A.R share. As a consequence of various research and development efforts, the concept of pig farming had changed from that of zero input enterprise to that of a semi-commercial one.

Tenth Five Year Plan (2002-2007)

Almost all the projects implemented during the 9th five year plan were continued in the tenth plan also. I.C.A.R launched the World Bank aided National Agricultural Innovation Project (NAIP). The basic components of NAIP were; research on production to consumption system, research on sustainable rural lively hood security and basic and strategic research in frontier areas of agricultural sciences. A grant of Rs. 20 crores was ear marked for N.R.C Pigs for initiating various research activities. To solve the acute shortage of boars for piggery development, the programme for proving exotic males in north eastern region was continued in the tenth plan.

Eleventh Five Year Plan (2007-2012)

As there was a reduction in pig population, which might be due to high slaughter rate and non availability of replacement stock, the goal for the eleventh five year plan was to achieve an overall growth of 6% to 7% per annum for the sector as a whole and for achieving a growth of 10% in meat production.

Twelfth Five year Plan (2012-2017)

The pig population again showed a reducing trend of -7.54% of the previous census. It might be due to the incidence of Swine Fever. Therefore in the 12th plan, in addition to other programmes, a Classical Swine Fever Control Programme was launched.

CHAPTER 2

SWINE BREEDING

In countries where swine production is highly organised, giant strides in the improvement of swine production have been achieved in the twentieth century. Major portion of this was due to improvement in the genetic material of swine through selection and breeding.

Economically Important Traits of Swine

The important objective of swine breeding is to maximise the efficiency of production. The traits which are most important with respect to economics of the enterprise are 1.) The number of pigs reared, which depends on a) Number of pigs born per litter b) Number of litters born per year. 2.) Economy of production, which depends on a) Growth rate b) Number of litters born per year and 3.) Carcass quality. Fancy traits like colour and traits of minor importance should be avoided.

Pigs have 19 pairs *i.*e. 38 chromosomes. Genetic selection within breed has produced strains of pig suitable for market requirement. All markets require minimum fat and maximum lean.

Litter size and viability: The number of pigs born per litter and the number that will be weaned out of them will determine the profitability of the enterprise. These characters are greatly influenced by management. Nevertheless, accurate estimates have shown that these traits do have a genetic basis too.

Economically Important Traits in Swine

Productive Traits

[Productive traits have moderate heritability, therefore improved by selection]

1. Mean litter size
2. Mean birth weight of piglets (kg)
3. Mean weaning weight
4. Mean weight at 48 weeks
5. Growth rate from birth to weaning
6. Rate of gain from weaning to market
7. Feed efficiency/kg of gain
8. Economy of gain
9. Milking ability

Reproductive Traits

[Reproductive traits have low heritability, therefore improved by cross breeding]

1. Number of piglets per litter (litter size)
2. Number of piglets weaned/litter
3. Prolificacy/fertility

Carcass & Structural Traits

[Carcass traits are highly heritable, therefore improved by selection]

Quantitative

1. Dressed weight
2. Dressing per cent
3. Carcass length
4. Back fat thickness
5. Loin eye area
6. Primal cut (%)
7. Meat to bone ratio

Qualitative

1. Shear force value
2. Fibre diameter
3. pH
4. Water holding capacity

Total litter weight at weaning is an aggregate measure of the pre-weaning performance in swine, as it is determined by the litter size, livability of pigs and the growth rate of pigs. It is also an indirect measure of the milking and mothering ability of the sow. Large litters born will not be of any use if the sow is a poor mother and crushes many of her piglets during their early life, or if the piglets are born weak or the sow does not have sufficient milk to feed them.

With the advent of early weaning systems, pigs can be weaned starting from three to five weeks instead of the standard eight weeks. In order to make selection easy, all pigs in a herd should be weaned at the same age. Even then, weighing of all the pigs in a herd may not be possible on the same day. Correction factors are available for converting pig weight to a standard eight weeks basis. One such adjustment is calculated by the formula:

$$\text{Adjusted Weight} = \text{Observed Actual Weight} \times \frac{41}{\text{Actual Age} - 15}$$

The formula has its greatest accuracy between 47 and 65 days of age beyond which the accuracy decreases.

Economy of gain from weaning to market: This has again two major components, i) daily gain from weaning to market, and (ii) efficiency of feed conversion. Daily gain from weaning to market influence profit because fast-growing animals have to be kept only for fewer days, thus requiring less labour, fewer days of housing and fewer days of exposure to the risk of becoming ill.

Weight at 154 days of age: includes both pre-and post-weaning growth and therefore, is a very important economic trait. To standardise pig weights to 154 days of age, from weights taken earlier or later, adjustment factors have been worked out as shown below.

$$\text{Adjusted Weight} = \frac{(\text{Actual Weight} + 154)}{\text{Actual Age} + 15} \times (154\text{-}99)$$

The efficiency of feed conversion (the amount of feed required per kilogram of gain by individual pigs) is difficult to calculate, as it requires individual feeding, which will involve additional cost. Due to economic reasons, it can be used only by swine evaluation stations and for evaluating boars in all herds. Besides, rate of gain and efficiency of feed conversion are related. It is not clear whether selection for rate of gain will automatically take care of feed efficiency. This will depend on the strength of relationship between the two traits, which varies with differences in breed, feeding and management.

Type and conformation score: Type and conformation are of value in determining potential production and carcass characteristics. The existence of relationship between conformation and production traits has been established. But, often these relationships are not strong enough to replace objective evaluation of carcass and production characters. Rapid changes can be brought about in the conformation of pigs through breeding. History of swine breeding in recent times reveals many changes in the fashions in swine type. To a certain extent, such changes had a bearing on the economic trends. But, often such changes far exceeded the limits of economic relevance and became fads with harmful consequences to the industry. For example, during the turn of the 20th century in United States, Poland China breed with a conformation- of refined, fat, short legged, type- a breed extremely small in build and of an early maturing nature, was preferred. This fad had its origin in the demand for lard as an export item from that country.

There were many fluctuations towards fattiness or leanness during the course of the century. At present, the objective is to produce pigs with minimum fat and maximum percentage of the primal cuts, namely, ham, trimmed loin, shoulder and bacon with desirable qualities of tenderness, juiciness and aroma. This is due to less demand for lard, as hydrogenated vegetable oils captured the market replacing lard because of the fear of cholesterol in animal fats causing arteriosclerosis, leading to heart troubles. Thus, a long, upstanding, lean pig is in demand at present.

Carcass trait: In concurrence with the consumer demand for more lean and less fat in pork, swine producers are paying more attention to carcass qualities. External appearance is only a very inadequate indicator of carcass quality. Therefore, pigs had to be slaughtered for carcass quality evaluations, rendering such information useless for breeding improvement with respect to those animals. But in recent years, three methods have been evolved to evaluate carcass quality in the live animals *viz.* the probe, lean meter and ultrasonics (Fig. 2.1). The probe consists of a sharp scalpel blade attached to a narrow metal scale, which is pushed through the fat down to the loin muscle. When it reaches the loin muscle, a firm resistance will be noted. A clip on the probe is then pushed down to the skin line. The probe is withdrawn and the measurement is noted. The lean meter is based on the difference in electrical conductivity of fat and muscle. In ultrasonic equipment, the time elapsed between the introduction and the return of reflected pulse (high frequency inaudible sound) is calibrated to measure the fat depth. Ultrasonic back fat reading is made at the last rib, usually on the midline of the back (Fig. 2.1). The ultrasonic equipment may also be used to estimate loin eye area at the last rib. But these methods do not give any idea about fat deposited in between muscles (marbling) or the quality of lean meat. Therefore, there is continued necessity for slaughter and carcass evaluation to provide proper direction in breeding programmes. Carcass data on littermates will be of great value in this regard. Carcass data may be collected from one male and one female in a litter.

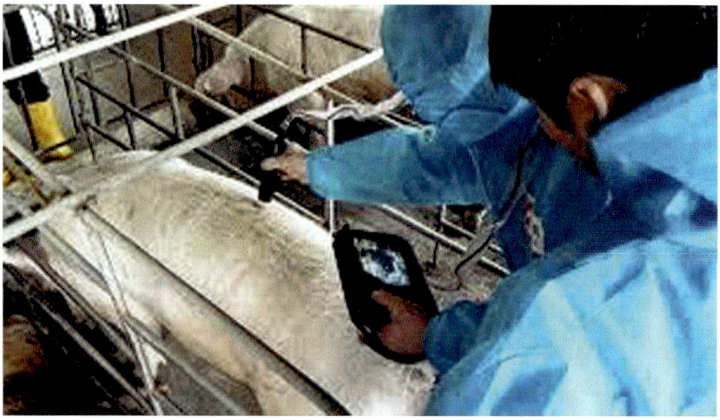

FIG 2.1 BACK FAT TESTING USING ULTRASONIC EQUIPMENT

Heritability of traits

The choice of an appropriate breeding system and selection procedure depends on the nature of genetic variability in the co-variability between the traits to be considered. The average and range of estimates of heritability of different traits of economic importance are presented in Table 2.1.

TABLE 2.1 HERITABILTIY ESTIMATES OF SWINE TRAITS (Biswas, D.K., 1972)

Traits	Heritability estimates	
	Average	Range
Performance Traits		
Litter size at birth	0.11	0.06 to 0.15
Litter size at 21 days	0.10	0.10
Litter size at weaning	0.11	0.09 to 0.12
Litter weight at weaning	0.19	0.17 to 0.21
Daily weight gain (Weaning at 90 kg)	0.32	0.28 to 0.39
Pig weight at 5-6 months	0.30	0.30
Feed conversion	0.36	0.31 to 0.40
Carcass traits		
Loin-eye-area	0.44	0.35 to 0.55
Loin length	0.46	
Back-fat-thickness	0.47	0.30 to 0.50
Carcass length	0.50	0.20 to 0.70
Belly thickness	0.48	0.43 to 0.52
Per cent ham in carcass	0.49	0.40 to 0.58
Per cent lean cuts	0.31	
Chemical fat in loin muscle	0.50	
Fatty tissue content in carcass	0.69	

[1] Several different methods were used by the various authors in calculating heritability.

[2] All averages are un-weighted means with respect to body-length, back fat-thickness, loin-eye area and percentage of lean cuts.

These estimates have been made from different stations, at different times with varying number of swine and by different methods. Understandably, a large variation is apparent between the estimates.

This might be due to real difference in the populations or their environments or simply due to different methods of estimation or even simply due to chance.

In spite of the differences between the estimates, one cannot help but be impressed by the tendency of the vast majority of the estimates to cluster within a fairly narrow range. This is evident not only for a particular trait, but also for specific kinds of traits. It can be seen that heritability is low for prolificacy and viability traits, intermediate for post-weaning weight, rate of gain and efficiency of feed utilisation and high for carcass traits.

Genetic relationships between traits

The estimates of genetic co-variability between the traits are subject to large sampling errors and it is also possible that the true relationships vary between breeds of pigs and presumably between different herds of the same breeds under different environments. It will be very difficult to have a precise estimate from any one single study. However, in general, it may be noted that most of the reported associations between major traits in swine are positive (favourable). One exception to this is the finding about rapid fat deposition and poor nursing ability. But, fortunately, these unfavourable relationships are not absolute. They may be slow, but not prevent, progress if selection is made for two or more unfavourably related traits.

Selection systems: It is expected that conventional selection techniques will bring about rapid improvement in traits with high heritability. This had been proved so, with respect to live animal back fat-thickness. Similarly, selection was successful in developing a high and low line with respect to carcass quality. These lines differed by 19 points in carcass score in a period of four generations of selection. Another striking example of efficiency of selection in swine traits may be found in Danish records for Landrace and Large White Yorkshire with respect to length, back-fat-thickness and belly thickness. Using conventional methods of selection, rapid improvement was possible in these traits.

Improvements in traits with intermediate heritability are expected to be possible through conventional selection methods, but slow. The view on the effectiveness of conventional selection scheme for improving intermediate heritability traits are far from uniform. There are controversial reports regarding the efficiency of conventional scheme for improving rate of gain or 154th -day body weight. But success has been achieved by some in developing high and low gaining lines as well as lines with high and low feed efficiency, through conventional selection techniques. Low heritability traits are not expected to respond to conventional selection techniques.

Actual experience shows that, intra-population selection is ineffective in improving the litter size and weight at birth and weaning. This is quite consistent with the low heritability of these traits. It might be that the fitness traits in general are subjected to unidirectional, natural and /or artificial selection, for a long time and hence expected to be low in heritability as well as unresponsive to intra-population selection.

Thus in general, conventional selection schemes have been effective for metric traits whereas their merit for intermediate and low heritable fitness traits is of doubtful nature. Major portion of the genetic variance for the latter group of traits are non-additive in nature, and is expected to yield to inter-population selection, with the aim to utilise both general and specific combining ability variance. A conclusive proof of the efficiency of such method is yet to come.

The country pig and the desirability of improving it

The indigenous pig is a nondescript scrub animal with very poor conformation and production traits. On the basis of limited works done, there does not appear to be any scope for large scale improvement of the country pig through selection or grading-up with exotic breeds. Country pigs do not possess any merit with respect to economic traits. Neither do they have any exceptional qualities of disease resistance or hardiness nor the ability to transmit such qualities to their progeny when crossed with exotic breeds.

Many times, it is seen that the purebred Yorkshire performed better than their crosses with country pigs, with respect to all important traits, including hardiness and disease resistance. Except the white colour of the Yorkshire, which is dominant over black, other characteristics of the country pigs remain largely dominant.

Coupled with their productivity, the stigma and sentiments attached to the country pig, make it totally unsuitable as the future animal in swine development. Therefore, wholesale introduction of exotic pigs was decided upon and exotic breeds of Yorkshire and Landrace have been imported and propagated in large numbers in state farms and by farmers. These breeds have already made a mark in the primordial swine industry of the country.

Mating systems

Under mating system, the relationship between the partners in mating to produce the progeny is considered. This can vary from close inbreeding, in which the sows and boars involved in the mating are close relatives such as brother and sisters, to the mating of unrelated animals in a breed and to cross breeding in which sows of one breed are crossed by boars of another breed. Many workers have investigated the effects of in-breeding. The observations in general, can be summarised as follows:

1. In-breeding causes a decrease in litter size at 1, 21, 56 and 154 days after farrowing, with the effects becoming progressively less, as the pig grows older. This means that the vigour of the pigs is adversely affected by inbreeding and that death loss before and after birth increases with the increase in the degree of in-breeding.
2. In-breeding has no effect on their growth rate up to 56 days of age. But at 154 days, there was 1.56 kg less weight per pig for every ten per cent increase in in-breeding.
3. In-bred sows are inferior in milking and mothering ability.

4. In-breeding delays onset of sexual maturity in gilts and boars. In-bred boars have less sexual libido.
5. In-bred gilts generally produce fewer eggs during oestrus and farrow smaller litters than those out bred.
6. In-breeding does not seem to have an adverse effect on economy of gains in swine. In fact, in some lines, it improves it.

However, out crossing, in which unrelated animals within the pure breed are mated has been practiced in many herds with excellent results. The excellent performance of purebred pigs of the Danish Landrace breed is an outstanding example. But, such results have been obtained under ideal management and environment. If the environmental and management conditions are not conducive for the full expression of their inherent abilities, often top class pigs with good hereditary endowment can perform badly.

To be successful, the commercial hog producer may have to follow a procedure of crossing in-bred lines or cross breeding. Crossing in-bred lines can lead to recovery of the vigour lost through in-breeding and at the same time keep the animals 'purebred'. The procedure lies in developing in-bred lines which are homozygous for many characters. They are then crossed to determine their combining ability for commercial production. But, the cost of developing in-bred lines and their poor performance has made this system not very popular.

It can be seen that heterosis or hybrid vigour will be best exploited by resorting to crossbreeding on those traits whose heritability estimates are low, as the non-additive gene effect is mainly due to dominance, over dominance and epistasis.

Heterosis is measured by the value, the crossbred offspring exceeds the average of the two parent breeds or in-bred lines for a particular trait.

$$\text{Per cent Hybrid vigour} = \frac{\text{Cross bred average-pure bred average}}{\text{Pure bred average}} \times 100$$

Traits high in heritability, like carcass length, back fat thickness, and loin eye area respond consistently to selection, but show little response to hybrid vigour. Traits low in heritability like litter size, litter weight, and survival rate usually demonstrate good response to hybrid vigour, which is revealed in Table 2.2.

TABLE 2.2 HERITABILITY AND TYPE OF GENE ACTION OF VARIOUS ECONOMIC TRAITS OF SWINE

Trait	Heritability	Type of gene action	
		Non-additive	Additive
Litter size at birth	0.11	Large	Small
No. of pigs weaned	0.11	Large	Small
Weight at weaning	0.19	Large	Small
Rate of gain from weaning to market	0.32	Moderate	Moderate
Economy of gain from weaning to market	0.36	Small	Large
Conformation	0.35	Small	Large
Carcass quality	0.44-0.59	Small	Large

Cross Breeding

Crossbreeding is being employed for commercial production of hogs more extensively than in any other species of livestock. Many experiments have been conducted to evaluate the effect of crossbreeding in swine. The general finding is that there are economically important advantages in favour of crossbreds. Experimental work has also shown that crossbred sows show heterosis with respect to mothering ability. Using crossbred sows, as mothers, gives the opportunity for the hog raised to take advantage of heterosis in the important traits, side by side with that of livability in pigs and rate and efficiency of gain. Besides, this will remove the necessity of having to start every time with purebred sows. Incorporating these ideas, systems of criss-crossing and triple or rotational crossbreeding have been developed. In criss-crossing, the crossbreds are bred back to the parental purebred breeds alternatively, which is represented below.

Purebred (100%) A x Pure bred (100%) B

↓

Crossbred A (50%) B (50%) x 100% A

↓

Crossbred A (75%) B (25%) x 100% B

↓

Crossbred A (37.5%) B (62.5%) x 100% A

↓

Crossbred A (68.75%) B (31.25%) x 100% B

↓

Crossbred A (34.37) B (65.63)

It can be seen that, a crisscrossing system will stabilise with the crossbreds having two-third of their inheritance from the breed of their immediate sire and one-third from the other breed used in the system.

Complementary

Complementary refers to the advantage of crossbred over another cross or over a purebred, due to the manner in which two or more characters combine or complement each other, the objective is to get desirable trait of each. Thus in crossbreeding programmes, breeds that complement each other should be selected, there by maximising the desirable traits and minimising the undesirable traits.

Rotational crossing

This is a breeding system which involves two or more breeds. The advantage of this system is that the heterosis level of the progeny is kept optimum by simple and flexible breeding operations.

Three breed rotational cross

By rotating the breeds without back crossing to one of the original breeds, high levels of hybrid vigours are maintained. The females born by crossing breed-I with breed-II are mated to male of breed-III. The female progeny born to such mating is crossed to boars of breed-I and the female progeny born inturn to such mating to breed-II and so on (Fig. 2.2). The crossbreds born will have $^4/_7$th of the inheritance of the breed of their immediate sire, $^2/_7$th from their maternal grandsires breed and $^1/_7$th from the third breed involved. Rotational crossing can be represented as follows.

Purebred A (100%) x Pure bred B (100%)

↓

Purebred C (100%) x Crossbred (A 50%, B 50%)

↓

Crossbred (A 25%, B25%, C 50%) x Purebred A (100%)

↓

Purebred B (100%) x Crossbred (A 62.5%, B12.5%, C 25%)

↓

Crossbred (A 31.25%, B 56.25%, C12.5%) x Purebred C (100%)

↓

Crossbred (A 15.6%, B 28.1%, C 56.2%)

In both these systems, the heterosis of the offsprings and the milk production capacity of the dams are exploited. Thus, the total production will be as great and greater than in first crossed. Extensive study has been carried out at the Iowa Agricultural Experimental Station to evaluate the comparative performance of purebreds and crossbreds, the results of which are summarised in Table 2.3.

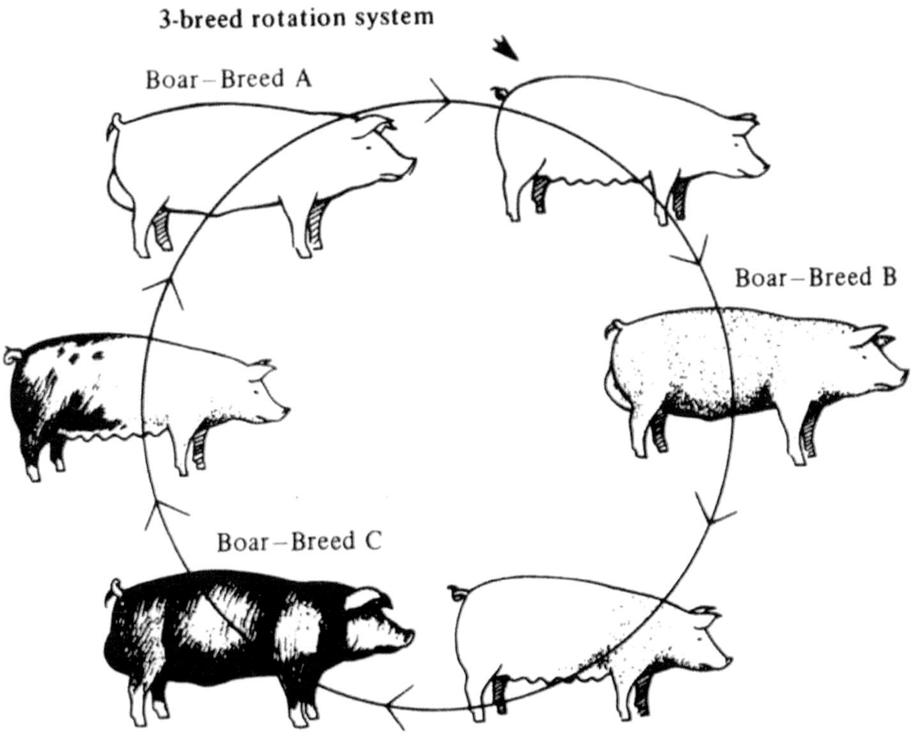

FIG. 2.2 THREE BREED ROTATIONAL CROSS SYSTEM

TABLE 2.3 PERFORMANCES OF CROSSBREED SWINE COMPARED TO PUREBREEDS

Trait	Pure breed	First cross	3-breed cross	4-breed cross
Litter size at birth	100	101	111	113
Litter size at 8 weeks	100	107	125	126
Weight at 8 weeks (per pig)	100	108	110	109
Weight at 154 weeks (per pig)	100	114	113	111
Pork production per litter	100	122	141	140

Note: All data expressed as percentages of purebred performance

Source: Hazel (1963).

Roto-terminal Cross System

This system produces maximum progeny heterosis and a good proportion of maternal heterosis. Through optimum positioning of the breeds, such a system makes for superior matching (complimentary) of the breeds involved and is conducive for within herd replacement that can reduce the cost of replacement stock and reduce the risk of disease.

This system is in between the rotational cross and terminal cross. In this system a sub-breeding herd is established. The replacement gilts are produced by two or three breed rotation in the sub-breeding herd. The gilts are then mated to a terminal sire *i.e.* boars from a specialised strain to produce the market hogs (Fig. 2.3).

FIG 2.3 ROTO TERMINAL CROSS SYSTEM

Terminal Cross System

Due to the genetic diversity that exists between pure breeds, this is becoming the preferred system of breeding in large units. This system will provide maximum hybrid vigour in the progeny and the dam, makes superior matching (complimentary) among the breeds utilised, and results in a final product of consistent performance and composition.

In this system superior F_1 gilts for maternal performance are produced by mating pure bred boar of one breed to pure bred boar of another breed. These F_1 gilts are then mated to terminal sire *i.e.* a boar belonging to a third breed that has superiority in growth and carcass trait (Fig. 2.4).

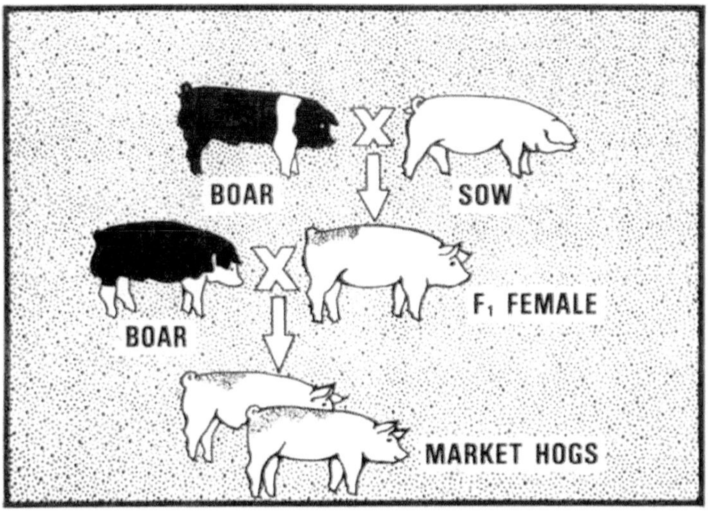

FIG 2.4 TERMINAL CROSS SYSTEM

Maximum performance by commercial swine breeds can be obtained through judicious crossing of different breeds of swine following a definite plan of breeding that suits the management system. There is no incompatibility between following a crossbreeding system and producing replacement sows.

Sows can be kept up to 3-pig crops and entirely replaced by gilts after sending the sows to the market. It is possible to replace all the sows in a herd by gilts from a single pig crop because of the high prolificacy of swine. Boars of particular pure-breeds can be purchased from seed farms.

Performance Testing in Swine

This system is based on the accurate evaluation of the economically important traits of swine to determine the estimated breeding value

(EBV) of one animal as compared to another breed, based on genetic characteristics. Each year, the breeding centres are required to send test litters of four pigs each, so as to average two pigs from each sow. The performance of pigs from various herds is tested under uniform management and feeding conditions to provide viable data comparisons. The world's first swine production testing station was established in 1907 in Denmark. This was followed by United States and Canada. The average daily gain of the Danish Landrace pigs was raised from 0.54 kg to 0.67 kg during the period from 1910 to 1963 through the efforts of Danish swine testing programme. The Danish and Canadian systems are based on progeny tests. In U.S, a comprehensive testing and genetic evaluation is carried out for the purebred swine industry. It is called the Swine Testing and Genetic Evaluation System (STAGES). The various traits evaluated are:

1. Litter size at birth (which has a heritability of 15%).

   The quantity of feed consumed by the sow between breeding and the date of weaning of her first litter is computed. On an average, the sow will have consumed a total of 1 tonne of feed during this period. Thus if the price of feed is worked against a litter size of 4 or 5 piglets, the eventual profit will be small.

2. Litter size at weaning (which has an approximate heritability of 12%).

   Litter survival to weaning is a measure of the mothering ability of sow, although it has bearing on the scientific management practices adopted by the caretaker of the heard.

3. Birth weight of pigs (which has a heritability of 5%).

   Piglets born with lower birth weight lacks vigour and usually gets crushed by the sow.

4. Litter weight at weaning (which has a heritability of 17%).

   Weaning weight is important because it has been shown that the pigs that are heaviest at weaning time reach market weight more quickly. The low heritability of this factor indicates that it is largely a function of the nursing ability of the sow rather than genetic.

5. Average daily gain from weaning to marketing (which has a heritability of 30%).

   Daily rate of gain from weaning to marketing is important because a) it is highly correlated with efficiency of gain and b) it takes a shorter time for reaching the market weight, thus saving labour, reducing risk and less exposure to disease. It allows for rapid turnover of capital in the shortest possible time.

6. Feed conversion efficiency (which has a heritability of 29%).

   Accurate litter feed records should be kept because the most profitable animals generally requires less feed to reach 120 kg.

7. Conformation score (which has a heritability of 29%).

   Low quality herds are likely to have higher heritability figures.

8. Carcass characteristics

   a. Length (which has a heritability of 60%).

      Carcass length is perhaps the most highly heritable trait in hogs. This accounts for the rapid shifts that have been observed frequently.

   b. Back fat thickness (which has a heritability of 50%).

      The probe, lean meter or ultrasonic equipment can be used to measure back fat thickness on prospective breeding animals.

   c. Loin eye area (which has a heritability of 50%).

      Loin eye area is an indication of muscling or red meat.

   d. Per cent ham, out of carcass weight (which has a heritability of 58%). Ham is high-priced cut; hence, the aim is to get a large ham as possible.

   e. Per cent lean cut, out of carcass weight (which has a heritability of 50%).

      A high yield of lean cuts denotes less fat and more edible meat.

Swine Selection

Selection is the judicious process of selecting the individual pig as parents for the next generation. Much of the improvement in swine production is a result of the strict scientific selection and breeding.

Selection of Boars

The boar influences conception rate, litter size and contribute to one half of the genetic makeup of his progeny. Boar selection is very important, because the boar becomes the parent of more offspring's than the female, as the normal boar: sow ratio is 1: 15 to 20 sows.

The following are the important criteria for the selection of boars for breeding purpose.

1. They must be offspring's of better producing gilts and sows and should be a good representative of the breed with pedigree and conformation.
2. They must belong to a litter with size and weight of not less than 8 and 72 kg respectively at weaning.
3. They should have signs of masculinity with good testicular development and at the same time should possess good temperament and docility.
4. Boar must possess traits of meaty hog *i.e.* body length, depth, height, skeletal size, muscle size and shape.
5. They should be free from physical defects and must possess soundness of feet, legs, udderline, body capacity and should have mating ability.
6. They should have high average daily gain and high feed conversion efficiency.

Boars satisfying the above stipulations are subjected to preliminary selection at 5 months of age, provided they attain a body weight of at least 60 kg. Final selection of boars should be undertaken at 7 months, provided it has attained a body weight of at least 90 kg. An irritable boar, difficult to drive and one which is inclined to fight, may transmit the nervous disposition to the piglets. This may make them poor mothers.

Selection of Sow/gilt

The sow productivity has improved very much in recent years due to careful selection and culling pressure. The sow also contributes half of the genetic material to its progeny and the mothering ability of the sow can make difference in raising viable piglets out of the total farrowed. The important criteria for the selection of sow and gilts are;

1. They should be selected from mother pigs having good mothering ability as shown by large litter size and weight at birth and weight at weaning of piglets.
2. They should have well developed udder with 12-14 functional teats and at least 6 teats in each row evenly distributed on belly side.
3. Teats of sow must be free from all abnormalities like inverted teats and blind teats.
4. They should have good mothering ability with quiet disposition.
5. The sow should have good reproductive soundness and normally developed reproductive external genitalia.
6. Select sows that are already bred at least once, with minimum back fat thickness.
7. The sow must be able to produce numerous viable young ones each year.
8. Sow must be ready for re-breeding towards the end of lactation.
9. Gilts should have sound feet and legs as it interferes with normal breeding, farrowing and nursing functions.
10. Select the fastest growing, leanest gilts, which are sound and from large litters.

Sows which fail to rebreed, which has small litters, failed to milk or having farrowing problems should be culled.

SELECTION SYSTEMS

Tandem Selection

In tandem selection, only one trait is selected at a time until maximum progress has been achieved; then another trait is selected for improvement. A good example of this method was the change from the lard-type hog to the long, lean variety. Rapid progress was made in these traits as the heritability for this trait was higher than most other traits. The limitation of this system is that since many traits are interrelated, the selection for one may not always result in improvement of another.

Independent Culling Level

Independent culling is a systematic procedure, where minimum standards are set for several traits. A failure to meet these standards by any individual will result in the culling and removal of that individual from the herd. This system works well and is most suited to swine because large litters give many numbers to work with and if the culling pressure is high, re-establishment of the herd is relatively rapid.

Selection Index

By using a selection index, all important traits can be easily combined into one figure or score. Higher scores mean more valuable animals for breeding purposes. The weight assigned to each trait will depend on its economic importance, heritability and genetic linkage to other traits.

Best Linear Unbiased Prediction (BLUP)

Best Linear Unbiased Prediction (BLUP) programmes are complex computer-based calculating systems, which utilise various mathematical models to allow accurate evaluation of sire for his breeding value and improve, greatly the accuracy of test and the construction of indices, especially for traits of low heritability such as those associated with reproduction. BLUP entails the distribution of large single nucleus population to be distributed over numerous herds and geographical sites, the concept of the group nucleus.

Genomic Selection

Genomic Selection (GS) has become a major breakthrough in animal breeding and genetics. Its application could considerably increase the genetic gain of economically important traits. The Swine Genome Sequencing Consortium (SGSC), a partnership of institutes involved in sequencing and genomics, completed the whole genome sequencing of pigs by December 2009. Over 6 lakh BAC end sequences were generated from four libraries in three laboratories. The end sequences are available at the National Centre for Biotechnology Information (NCBI), trace archives. Previously, markers corresponding to the Quantitative Trait Loci (QTL) specific for important traits were only identified and used for selection known as Marker Assisted Selection (MAS). Now markers associated with the QTL responsible for the expression of important traits had been identified in the whole genome sequence by Genome wide association study (GWAS) and selection procedure is carried out by utilising the data of whole genome marker which is known as genomic selection. The development and use of Single Nucleotide Polymorphism (SNP) chip of low density for genomic selection in pigs has made genomic selection applications economically feasible. Genomic selection has resulted in the accuracy of breeding value over traditional selection (BLUP) by 68%. It also opens up windows to new selection scheme design and evaluation of novel phenotype. One of the most important advantage in genomic selection model is that it can accommodate non additive effects which are valuable in low heritability traits like litter size. The limitations of genomic selection are the generation lag between crossbreeding generation and selected candidates and the very high cost of genotyping.

Culling of Boars

20% of the breeding boars are annually culled and replaced with young stock. The parameters for culling are: Boars of over 5 years of age, those with irritable nature and nervous disposition, those not true to the breed, infertile ones, those which are over fat and too heavy, those with weak limbs and cryptorchids.

Culling of Sows

One third of the older sows are annually culled. Gilts and sows not settled after 4 breeding period, nervous and irritable ones, those which produce small litter, sows with defective teat like inverted and blind teats and poor milkers, and those with small external genitalia are culled. Gilts and sows which do not meet the standards of meaty hogs are also culled.

BREEDING MANAGEMENT

Farrowing Systems

The different farrowing systems are one litter a year system; two litter a year system and multiple farrowing systems.

One litter a year system

One litter a year consist of pigs that are maintained throughout the year to raise one litter of piglets. One litter system is based entirely on the use of gilts that are finished and marketed soon after weaning their litters.

Advantages of one litter system are:

1. It is easier to stick to the schedules and management problems are less.
2. Total capital investment is less on building and equipments.
3. Less feed storage is required.
4. Farrowing can be planned during favourable climate.

Disadvantages of one litter system are:

1. Building and equipment are not used to the maximum.
2. Improvement in the herd is limited as there is only one litter per year and as the sows are sold after one litter.
3. The labour requirement is not distributed throughout the year.
4. Farm income is not distributed throughout the year.

Two litter a year system

In two litters a year system, the breeding programme is managed in such a way that the sows are made to farrow twice in a year.

The advantages of two litters a year systems are:

1. Maximum use of the facilities created by capital investment is made.
2. Rapid improvement of the breeding herd can be effected.
3. Better distribution of labour.
4. Pigs are marketed two times a year.
5. Better distribution of farm income.

The disadvantages are:

1. More loss in young ones as farrowing takes place during inclement weather.
2. More expense towards concentrate feed, protein supplements etc.
3. Pigs being thrifty have lower rate of gain.

Multiple Farrowing system

Multiple farrowing refers to a breeding programme in which, there is a strict scheduling of breeding so that litters are produced throughout the year. In practice the sow herd is split into either 2 or 3 groups with each group farrowing twice a year. If two groups of sows are used, pigs farrow every three month. If three groups of sow are used, pigs farrow every other month.

Advantages of multiple farrowing systems are:

1. It keeps the swine market steady and stable without any price fluctuations.
2. It distributes work load for the swine producers.
3. It makes maximum use of the existing building and equipment.
4. It provides a steady supply of pork throughout the year.
5. It avoids sharp price rise.

The disadvantages of this system are:

1. More experience and expert consultation is required to manage the prolonged farrowing season.
2. Increased possibility of disease outbreak because of the likely built up of pathogenic organisms.

All-in/all-out system: In all-in/all-out system, group of pigs farrowed within a short period of time (within two to three weeks) are reared in groups from growing to finishing and then sold out together. The incidence of disease in the herd is reduced in this management method; it also improves the rate of gain and feed efficiency.

Age of Puberty

Boars: Although boars reach sexual maturity at 6-8 months of age, it is generally recommended that they must be at least 10-12 months (100 kg) before being put to regular use. Sexual behaviour of males has been described by Houpt. (2012). It includes phases of courtship, mounting, intromission and ejaculation. The duration of copulation is 3 to 20 minutes. Sex pheromones synthesised in the testes of the boar are transferred to the salivary glands and are secreted during pre-mating behaviour which enhances the receptivity of the female to the boar. 5-androst-6-en-3ol is one such hormone.

Gilts: The onset of puberty in female pigs occurs at 5-7 months of age, but they are put for breeding only after 8-10 months at about 80-100 kg. The number of ova released per oestrous increases gradually over the first several oestrous cycle. Rather than age, proper development of gilt is essential to withstand the strain of lactation.

The factors which affect the age of puberty are breed, sex and environment, especially nutrition.

Oestrous Cycle

The length of oestrous cycle in gilts and sows varies from 18-24 days with an average of 21 days. The cycle can be dived into pro-oestrous, oestrous and dioestrous. During pro-oestrous (1-3) days, females are alert of the approach of the boar, will mount other females, accept mounting by dioestrous females, but will not tolerate mounting by a boar. During oestrous period, the female will permit mounting only by the boar. The external signs of oestrous are; restlessness, grunting, chomping of jaws, swollen and reddened vulva, vaginal discharge, mounting of other animals, frequent mounting on other animals, frequent urination and 'standing heat reflex'. The 'standing heat reflex' consists of immobilisation, arching of the back and in some breeds the cocking of the ears. This reflex is normally exhibited by gilts/sows in heat in the presence of boar; but may also be elicited by applying downward pressure on the back of the animal called the 'riding or back pressure test' (Fig. 2.5). The ideal heat detection system would involve the stock person performing the riding test in the presence of a boar because the boars release the necessary stimuli to induce this reflex. Heat detection can also be efficiently carried out using a vasectomised boar called the 'Teaser'. Though the oestrous period may last for a period of 3-4 days, the female will mate only for a period of 2-3 days. Gilts will cycle earlier if a boar is placed in adjoining lot allowing the gilt to see and smell the boar. Allowing sows to have fence line contact with boars also stimulate oestrous.

FIG. 2.5 'RIDING TEST' OR 'BACK PRESSURE TEST' FOR HEAT DETECTION
Source: Dr. V. Ramesh, V.C.R.I, Namakkal

Ovulation

Ovulation occurs late in oestrous, usually on the second day after the onset of oestrous. The time span for all ova to be released in a given oestrous may be up to seven hours. As age advances, the number of ova released during a single oestrous increases from 8-10 in the first post pubertal heat to 12-14 at the third heat. Sow ovulates about 15-20 eggs per oestrous. Some sow's ova are never fertilised; others that are fertilised die in early development. Ova released early are fertilised sooner and may have lower embryonic mortality than those released towards the end of seven hour period.

Plane of nutrition especially before breeding may affect the number of ova. Giving additional feed at least seven days before the next anticipatory heat results in the increase of about 2-3 ova per ovulation. Short term feeding, 24 hours prior to breeding had no effect on the process of folliculogenesis and ovulation. Environmental temperature also influences ovulation rate. Increased exposure to a high environmental temperature (>30 °C) is detrimental for the survival of ova and also results in high embryonic mortality.

Time for mating

For maximum fertilization of the ova, mating is usually done 24 hours after the onset of heat symptoms, a second mating is advised 8-12 hours after the first mating. This is done because the ova are released on the second day of heat and sperms must be present to fertilize the ova before they die (within 12 hours of ovulation). This practice results in more number of ova being fertilised, higher conception rates and thus large litters. Gilts are bred when they are 7 to 8 months of age and weighs about 100-120 kg. Gilts produce large litters if they are bred during their second heat period, rather than during their first heat period. Check for standing heat twice daily, gilts are bred at least twice at 12 hour interval and sows at least twice at 24 hour interval after standing heat is observed.

Mating Techniques

Hand mating: In this method, the sow in standing heat is brought to the individual boar for service in the boar sty. This is the most common method of mating in pure bred herds. In this method the farrowing rate is more accurate and the repeat breeders and sterile sow can be detected easily and culled regularly. When the temperature exceeds 27 °C hand mating should be carried out early in the morning or late in the evening.

Pen Mating: In this method, one boar is allowed to a group of sows (8-10) for detecting heat and mating. Maintenance of record is difficult in this method and there is chance for over usage of boar.

Flock or herd mating: In this method, the boar is permitted to run with the herd. Mature boar can usually satisfactorily serve 15-20 sows in pasture mating system. 2-3% of females will not conceive and will again come to heat within 21-25 days.

Fertilisation

Fertilisation occurs within 6-10 hours after natural mating. It is the process of penetration of the outer covering of the ovum (Zona pellucida) by the sperm. Normally, a single sperm enters the ovum but occasionally poly spermy does occur, especially if ovulation has taken place considerably earlier. Ova remain fertile as long as 24 hours after release.

Embryonic migration

Intrauterine migration of fertilised zygotes occurs in sow. The embryo enters the uterus at four cell stage approximately 48 hours after ovulation. About 6 days after ovulation the embryo reaches the stage of blastocyst. These embryos undergo intrauterine distribution in both the horns of the uterus, especially during 9th to 11th day. By the 13th day, the embryos are evenly distributed into both the horns of the uterus (Fig. 2.6). This mechanism is or optimum spacing of embryos for maximum uterine space and development.

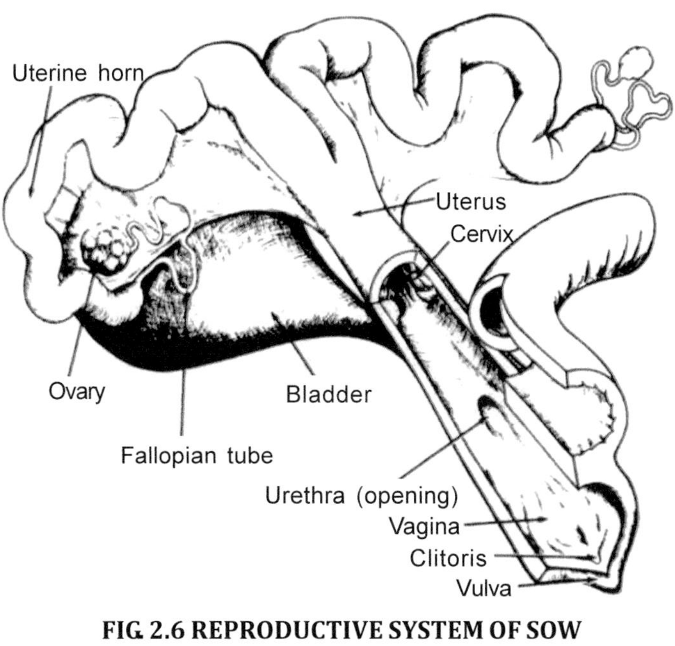

FIG. 2.6 REPRODUCTIVE SYSTEM OF SOW

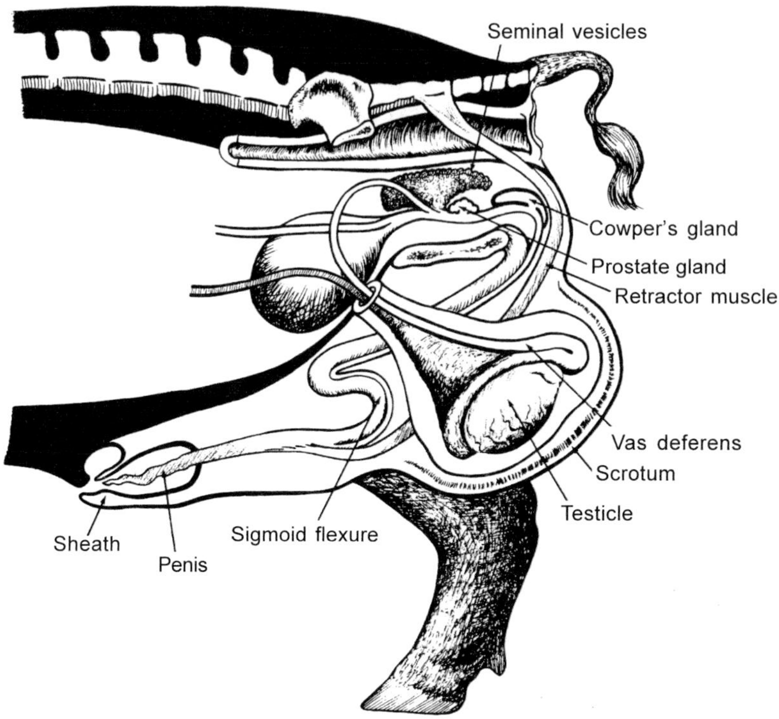

FIG.2.7 REPRODUCTIVE SYSTEM OF BOAR

Boar Semen Characteristics

The boar ejaculates between 150-500 ml of semen over a period of 1 to 15 minutes. This large volume contains about 40 x 10^9 to 80 x 10^9 spermatozoa. Ejaculation in boar occurs in three phases. 1.) A pre-sperm fraction consisting of 5-15 ml of a watery fluid containing few sperms. 2.) A sperm rich fraction, which is milky white solution *i.e.* 10-20% of the ejaculate volume containing up to 90% sperms and 3.) A post sperm fraction containing about 60-80% of the ejaculate volume. A gel fraction is expelled from the bulbouretharl glands at intervals (Fig. 2.7).

TABLE 2.4 IMPORTANT CHARACTERISTICS OF PIG SEMEN

Ejaculate Characteristics	Normal Value
Ejaculate Volume, ml	100-500
Total Sperm Number x 10^9	10-100
Motility %	70-90

The main assessment of pig semen characteristics is based on sperm numbers, motility and morphology, which reflect the functions of the seminiferous epithelium and epididymal sperm maturation. The colour of the sperm rich fraction should be creamy white. Normal values of important semen characteristics are given in Table 2.4.

Motility is expressed as the percentage of the sperm which are actually progressively moving compared with the total number of sperm present. Expected values are 70-90%. Sperm number is most accurately measured by counting sperm cells on haemocytometer, or assessed more quickly by measuring cloudiness using a calibrated colourimeter. Sperm morphology should be assessed using high powered magnification in order to measure the amount of structural abnormality.

Artificial Insemination (A.I)

Two primary reasons for using A.I. in swine are 1.) Disease control and 2.) Genetic improvement. A.I allows for the development of a close herd. It allows introduction of new genetic material with minimum risk. It allows better use of superior boars. The disadvantages of A.I in swine are the high level of technical input and labour requirement. Without properly trained and experienced staff, the conception rate and litter size may go down with A.I than natural service. Hotter climates are least appropriate to A.I operations as acute heat stress reduces total number of sperm per ejaculate, reduces sperm motility, increases abnormal sperm per ejaculate and reduces embryo survival.

A boar may ejaculate 50×10^9 sperms at any one time but only 2×10^9 progressive motile sperms are only required for effective fertilisation. Therefore, one ejaculate can be used for 25 or more mating through the practice of A.I. A mature boar should not breed more than 2 or 4 females per day. A.I makes it possible to breed at least 10 or more sows in one day from one ejaculate. For this the semen of the boar has to be collected, diluted and then used to inseminate those female in heat that day.

Boars can be trained to mount a suitable 'artificial sow' (Fig. 2.8) for semen collection in about 14 days. Collection can be carried out every 3-4 days as this gives the best sperm concentrations. Semen is collected into an appropriate container (at body temperature) using a gloved hand to serve as a surrogate cervix. The sperm free fraction is discarded and the sperm rich fraction is collected. This is filtered of accessory fluids and held at 30 °C.

A volume of 100 ml per ejaculation can be expected. Following microscopic examination for sperm density, sperm motility and sperm morphology, the semen is then diluted with nutrients (to maintain sperm activity for a reasonable period of time), buffers (to maintain the proper biochemical medium for sperm viability), chilling media (to protect the sperm from cold shock) and antibiotics (for disease prevention). Thus each ejaculate can be made up into 25 individual doses of 50 ml each with diluents to contain about 2 x 10^9 sperms per dose. The diluted semen can remain potent for 3-5 days at 20 °C. For best results two inseminations should be carried out at an interval of 16 hours for maximum efficiency. The artificial insemination catheter should be securely locked into the cervix of the sow and the 50 ml dose is allowed to flow slowly at its own rate into the cervix and into the uterus of the sow with minimum back flow and loss of semen.

FIG. 2.8 ARTIFICIAL MOUNT USED FOR SEMEN COLLECTION FROM BOAR

Though chilled semen technology has been successfully used for swine A.I, the frozen semen technology is yet to prove commercially viable. The main reason for this are: 1) The citrate and phosphate extenders which are capable of prolonging the life of bull sperm, are toxic to boar sperm under the same condition. 2) Boar spermatozoa are subjected to "cold shock" when cooled suddenly to 6 °C or less. 3) Addition of substance such as glycerol and dimethyl sulfoxide (DMSO) to the diluent to prevent the loss of potassium and other ions has been unsuccessful. 4) Boar semen exhibits anabiosis – that is, quick loss of motility and settling out on short-term storage.

Transgenic pigs

Transgenic pigs are created by mico-injection of DNA into the pro-nucleus of the ovum, where the transgene may replace or be in tandem with the resident gene at a single chromosomal site. Transgenic pigs have already been created and shown to contain genes from other species, but there are many problems before the genes are fixed and express positive benefits, generation to generation. Integration of genes into chromosome is random and effective transfer requires close targeting. Targeting assumes detailed knowledge of the gene map. Gene transfer remains as the most fruitful way of using genetic engineering to improve growth, lactational and reproductive traits through transgenic manipulation of physiological process.

Embryo Transfer and Cloning

Swine are naturally polytocus animals producing multiple ova for multiple births. In pigs, the embryo transfer and cloning technique can be used to increase the number of progeny obtained from genetically superior females, by using common and inferior recipient mother to incubate the developing foetuses. This technique can also be used to transfer genes into nucleus herds of high health status by using fertilised embryos thereby, avoiding transmission of diseases. The flushed fertilised embryo can be divided by micromanipulation of nuclei and in this way clones can be created.

The best genes available from the dam and sire could be combined in an embryo, which can then be subdivided to make many identical highly superior individuals. With a combination of cloning and embryo transfer, sire lines can be produced from dams of high uterine capacity. This will increase selection rate within the line, reduce cost and increase the rate of proliferation of the line.

TABLE 2.5 NORMS FOR NORMAL SWINE REPRODUCTION

Age at puberty (boars)	6-8 months
Age at puberty (gilts)	5-7 months
Breeding age of boars: Approximate body weight	1-1$^1/_2$ years, 80-110 kg
Breeding age of gilts: Approximate body weight	8-10 months, 80-100 kg
Sex ratio	1: 20
Oestrous cycle	18-24 days average 21 days
Oestrous period	2-3 days
Optimum time for mating	24 hours after the onset of heat signs second mating is advised 8-12 hours after the first mating
Number of service per conception	Two at an interval of 12 hours
Gestation period	114 days (3 months, 3 weeks, 3 days)
Suckling period (weaning)	56 days
Average litter size at birth	10-12
Average litter weight at birth	12-14 kg
Birth weight	1.2-1.3 kg
Average litter size at weaning	8-10
Average litter weight at weaning	120 kg in gilts and 150 kg in sows
Weaning weight	12-13 kg
Occurrence of post weaning heat	4-7 days
Volume of ejaculate	200 ml
Average number of sperms per ml	1,00,000
Re-mating of sows	1st post farrowing heat
Average age for castration of piglets	2-3 weeks
Market age of fattening pigs	7-9 months
Weight at 6 months	70-75 kg
Market weight	100-120 kg
Farrowing interval	7-7$^1/_2$ months
Farrowing rate	1.8-1.9
Average life of sow	Up to 6 litters

Pregnancy

The gestation period of the sow is generally recognised as 114 ± 2 days (3 months, 3 weeks and 3 days). However, individual sows may have normal gestation period that vary over a wide range, therefore, both good records and close observation are essential to ensure that the sow is at the right place at the right time for farrowing. The most critical times during gestation are the first 30 to 35 days and just prior to farrowing. Most embryonic death occurs during the first 30 to 35 days of gestation. The peculiarities of foetal development in swine are; multiple ovulations and fertilised multiple embryos develop in separate embryonic membranes. Gilts and sows should not be allowed to become too fat during gestation period. A gain of 22-34 kg for sows and 32-45 kg for gilts is the right body weight gain for sows and gilts. Experiments at Indian Veterinary Research Institute (I.V.R.I), Izatnagar have shown that 2-2.5 kg of concentrate feed per day is sufficient for pregnant sows.

Pregnancy diagnosis

The physical activity of the female pig will be reduced and it becomes more docile during pregnancy. The development of mammary system is another indication of pregnancy. Normally, gilts will show some signs of udder development after 2 months of pregnancy. There will be change in texture and development of udder 2 to 3 weeks before farrowing.

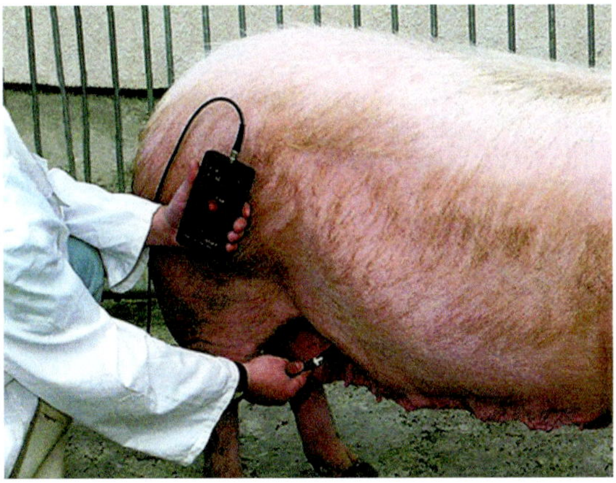

FIG 2.9 ULTRASOUND TESTING OF PREGNANCY

Pregnancy diagnosis is carried out with 90 to 95% accuracy in swine using ultrasonic detectors (Fig. 2.9). They give accurate and best results when used between 30 to 45 days after mating. The Uterine fluid, which increases rapidly following conception, reach detectable levels, 30 days after breeding. The high frequency inaudible sound emitted by the transducer is echoed back by this fluid, which is detected by another part of the transducer. This method can be used to detect pregnancy up to 80 to 90 days after breeding following which the mass in the uterus exceeds the fluid content.

Signs of farrowing

The sow demonstrate several signs of impending parturition these are:

1. Change in the texture of the mammary gland: The mammary gland changes from a soft flabby structure to one of increased firmness and turgidity as parturition nears. These changes usually becomes noticeable several days prior to parturition.
2. Availability of milk: A serous secretion may be expelled from the mammary glands of some sows as much as 48 hours prior to delivery. This secretion changes to a milky consistency near delivery. When milk becomes readily available, one can expect the first pig within 6 to 8 hours.
3. Bed making and restlessness: As parturition approaches, there is an overall increase in activity on the part of the sow. She gets up and lies down frequently and may attempt to collect things within her reach to make a nest. She may also paw at the floor, chew on anything available and becomes more vocal. Urination and defecation may increase in frequency.
4. Vulvar swelling and discharges: Swelling of the vulva becomes obvious 3 to 4 days prior to delivery. The vulva becomes moist and blood-stained fluids may be observed within about 2 hours of farrowing. Small amount of meconium are often expelled prior to the first pig. When this is observed, the first pig will usually arrive within 30 minutes.
5. Respiration, pulse and rectal temperature all tend to increase as farrowing approaches and the general activity of sow increases.

Stages of farrowing

The farrowing comprises of three stages.

Stage – I Dilation phase of cervix

Stage – II Phase of foetus expulsion

Stage – III Phase of foetal membrane expulsion

Complete farrowing takes place in about 2 to 4 hours but in exceptional cases it may go up to 6 hours. The interval between births of piglets is 12 to 20 minutes.

Within two hours after farrowing the entire foetal membranes are expelled one after another. This must be disposed of efficiently so as to prevent the sow eating her foetal membranes.

About 30% of piglets are usually born in posterior presentation. Baby pigs are usually delivered with the sow in lateral recumbency. Sow tends to appear more nervous at the time of delivery of the early piglets. It is at this time that the greatest danger from crushing and mutilation exist. After the last piglet has been born, a nervous sow will usually become relaxed and settle down to nurse the piglets.

Baby pigs are on their feet within minutes after birth. On an average, they will have nursed successfully within 45 minutes following delivery. Occasionally, pigs may be born partially or completely enveloped in the afterbirth. The majority of these will struggle free with no negative after effects. In most instances, the umbilical cord usually breaks within 5 minutes after birth. Sometimes baby pigs are seen nursing with the cord still attached.

One per cent of the sows or gilts farrowing, require assistance with delivery. The following signs may be indicative of a problem.

- Abdominal straining and vulvar discharge without delivery of piglets
- Cessation of abdominal straining following delivery of only one or two piglets

- An interval greater than one hour passing since the delivery of the previous piglet with continued abdominal straining on the part of the sow.
- Presence of a foul smelling, coloured vaginal discharge.
- Anorectic, depressed, ill-appearing sow.

5-10 IU of oxytocin may be administered intravenously to induce contraction of uterine muscles for the expulsion of the foetus. If required injectable calcium preparations can also be given.

Normal Piglets and Still Births

About 5-10% of fully normal piglets are farrowed dead or still born. Their lungs do not float in water. Still births have been classified into two types. Type -I or Prepartum deaths (dead before the onset of parturition), Type – II or Intrapartum death (dying during parturition process)

Prepartum deaths

These deaths occur before the end of gestation. Infection during pregnancy (bacterial, viral and that associated with SMEDI syndrome) are responsible for prepartum deaths. Iron deficiency in the sow has also been associated with high incidence of still births.

Intrapartum deaths

These deaths occur during parturition due to non-specific causes influenced by duration of farrowing, litter size, position of birth order (Those piglets at the cervical end are born before those in the ovarian end). Those at the ovarian end have to travel 1.5 to 1.8 m of uterine horn during birth. Other causes are anoxia, anaemia, heat stress etc. 70 to 90% of all still births are of type-II and foetal anoxia during parturition has been recognised as the major cause. Such anoxia is induced by decreased placental blood flow associated with uterine contraction or with occlusion of the umbilical cord or premature detachment of the placenta. The incidence of intrapartum death is more in the second half of the birth. 70% being the last 3 piglets born.

Lactation

The baby pig's ultimate well being largely depends upon its ability to successfully attain adequate nutrition shortly after birth. The sow's antibody does not cross the placenta to the unborn piglets making adequate colostral intake during the first few hours essential for the survival of the new born piglets. To maximise the baby pigs opportunity for adequate nutrition, it is desirable that 14-16 functional teats be present. Teat abnormalities such as blind and inverted teats should be avoided and even spacing of the teat along the udder is desired. Adequate exposure of the lower row of teats during nursing is frequently a problem especially because the rear most teats in older sow have pendulous udder. Selection of gilts with udderlines closer to the midline may be helpful. The sow reaches maximum milk production in three weeks time, after that milk production declines.

Nursing

The normal nursing pattern consists of an uniform time interval of somewhat less than one hour between each feeding and this pattern continues over a 24 hours period. The three phases of suckling are:

Phase-I: - Comprises piglets jostling for the position on the udder. This phase lasts for 20-60 seconds.

Phase-II: - Nosing and butting the mammae. It takes about 30-40 seconds and oxytocin production is stimulated.

Phase-III: - Piglets become quiet for 20 seconds with slow mouth movement. Milk flow lasts for a mere 10-20 seconds and is characterised by rapid mouthing movements, while the piglets stretch their necks and flatten their ears and swallow 40-80 ml of milk. The sow herself during this period keeps grunting in a characteristic rhythmic vocalisation pattern of about one grunt every second. It is estimated that a sow produces about 4.55 kg milk per day. Colostrum is secreted during the first 2-3 days and the composition of colostrum is considerably different from that of milk. Colostrum is higher in percentage of total solids and protein than milk, but lower in ash, fat and lactose. The higher protein content of colostrums is due to its high immuno-globulin content.

Teat Order

Immediately following birth, fierce competition occurs among piglets of a litter for the mother's available teat. The teat order represents the dominance hierarchy by which the members of the litter may come to a general agreement as to which nipple they will suck. Teat order is usually established within three days of farrowing. Once the teat order is established, a given piglet usually maintains his chosen teat until weaning (Fig. 2.10). The anterior teats are most favoured and won by the bigger piglets. The anterior teats of sow produce the maximum amount of milk than the posterior once. In large litters, the weakest piglets may starve because they cannot compete for a nipple. Thus the runts or small pigs often perish as they are forced to the posterior, non functional teats. This is overcome by split-weaning, or fractionated weaning which disturbs the teat order (Page 147).

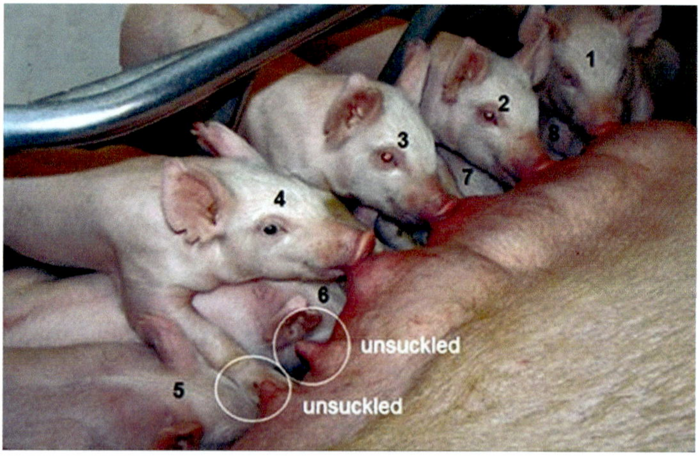

FIG. 2.10 TEAT ORDER IN PIGS

CHAPTER 3

SWINE FEEDING

Peculiarities of Swine Nutrition

Feeding accounts to over 75 per cent of the total cost of raising pigs. This may vary according to the type of stock being raised. A study of the cost and returns for 91 producers covering 4059 feeder pigs, conducted by the Agricultural Experiment Station of Arkansas University showed that feed costed 67.2 per cent. Fixed cost came to 17.2 per cent and other variables 15.6 per cent. Since the prices of feeds are higher and labour cheaper in India, percentage cost of feed will be higher and can safely be assumed to be above 75 per cent. Therefore, it is very important that the best use is made of feeds at the most economic level. Pigs consume more feed per unit body weight than do lambs, chicken or calves. They are also more efficient in converting feed than other species. This results in faster gain per unit body weight.

Pigs being an omnivorous monogastric animal, can digest only smaller quantity of crude fibre than ruminants. Therefore, their capacity to utilise roughage is limited. However, this does not mean that they do not need any roughage (Fig. 3.1). Far from it, providing good pasture or good quality hay is a nutritional necessity as well as it reduces feeding costs considerably.

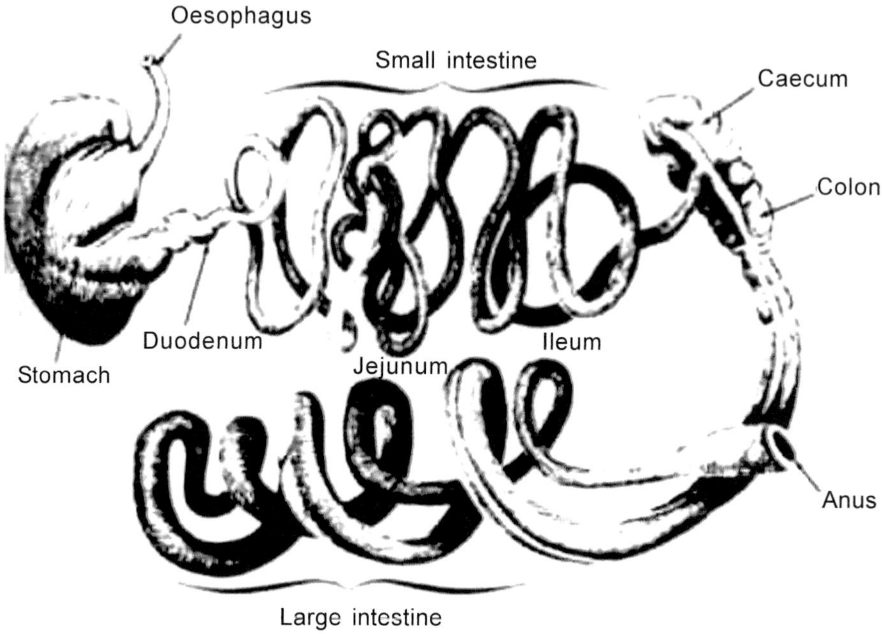

FIG. 3.1 DIGESTIVE SYSTEM OF PIGS

As different from ruminants, the amino acid content (protein quality) and the concentration of vitamins in the feed are important for pigs. In formulating rations for pigs, not only the total protein content but also the concentration of all essential amino acids has to be taken into account. Similarly, the ration should contain adequate amount of B-complex vitamin series like riboflavin, niacin, pantothenic acid and vitamin B_{12} apart from vitamins A and D.

The type of feed and the methods of feeding greatly influence feed efficiency, growth (rate of gain), breeding efficiency, carcass quality and the health in general. Needless to say, feed is the most important component that influences the profit from a swine enterprise.

Grain feeding: In all countries where swine production is well developed, grains form the basis for feeding; protein-rich concentrates and pasture or lucerne meal are supplemented. These grains like maize, oats, sorghum etc. make up the largest part of the feed. But, the question is whether a country like India, can afford to feed its swine on grains? Perhaps it may be possible to some extent in surplus states like Haryana, Punjab and Andhra Pradesh. But, by and large, states with deficient grain production should look towards pasture, lucerne hay, by-products of plant and animal origin, subsidiary crops like tapioca (cassava) and kitchen waste for feeding their swine. Utilising grains more than the required minimum for swine production will not only be uneconomical, but also criminal in countries where human beings are living below poverty line.

Pasture in swine feeding: Compared to dry lots, pasture is very economical in swine raising. Though pasture cannot provide a major portion of the nutrients required by pigs, it has the following advantages.

1. It makes feeding cheaper thereby increasing profits.
2. Provides all the vitamins.
3. Provides a portion of the protein requirement of the animal.
4. Provides adequate amount of minerals like calcium and iron.
5. Results in healthier pigs.
6. Generally faster gains are seen on pasture and
7. Saves grains for human consumption.

Lucerne pastures are considered the best for pigs. Red clover, sudan grass, ladino clover etc. also make good pasture for them. Usually, self-feeding methods are employed on pasture. Limited feeding will involve more labour, but will result in more efficient utilisation of the pasture. Pigs on pasture must be provided enough shade and adequate amount of clean drinking water.

Feeding Lucerne hay: In areas where pasture is not economical due to scarcity of land, dry lot feeding supplemented with Lucerne hay is a useful feeding practice. This may be fed as hay or as a meal. Lucerne hay may be self-fed from feeders. Legume hay should form 10 per cent of the ration of growing pigs and 20-35 per cent of that of sow; according to Washington Bulletin 296. This will have most of the advantages listed above.

Use of by-products and waste products: By-products of oil extraction industry (oil cakes), milling industry (wheat bran, rice bran, etc), dairy industry (whey, skimmed milk), slaughter-house (tankage, meat meal, chicken slaughter waste, bone meal) and fishing industry (fish meal) form good sources for swine feeds.

Garbage is widely used as a cheap swine feed. The only difficulty is that raw garbage sometimes spreads diseases like vesicular exanthema. Therefore, garbage should be cooked before feeding to swine. Cooking, however, makes garbage costlier and somewhat less palatable. Many human foods like wheat, pulse, etc., which have been damaged or contaminated, can sometimes be used as swine feed. Culled or surplus potatoes, yams, sweet potatoes, molasses, etc. can partially replace grains in swine ration. But they are not used to any great extent in this country.

Tapioca chips, tapioca meal, rubber seed cake and sal seed could be used to replace some of the more conventional feeds like maize, wheat bran and rice bran, even though a growth equivalent to that by feeding maize as a chief energy constituent of the ration cannot be achieved. In the state of Kerala and some other parts of the country, tapioca chips have already become an established swine feed. The nutritional requirements of swine vary with sex, age and physiological status of the animal. According to the nutritional requirements, swine can be classified into eight groups as follows.

Young pigs

1. Pre-starter pigs (2-5 kg)
2. Starter pigs (5-15 kg)

Weaners

3. Grower pigs (15-35 kg)
4. Growing and finishing hogs (35-90 kg)

Adult males

5. Boars

Adult females

6. Gilts
7. Pregnant sows and gilts
8. Lactating sows

Feeding Young Pigs

Make sure piglings get colostrum. The first milk of the sow is very important for the survival and future performance of the piglings.

Colostrum is rich in antibodies, vitamins and proteins, which will help in getting the pig started well. It has to be ensured that the piglets get their first feeding within one hour of farrowing for better passive immunity due to the phenomenon of 'gut closure'.

Feeding pre-starter pigs. It is the first feed for the piglets weaned before 3 weeks of age and is supposed to replace milk, known as milk replacer. Feeding pre-starter rations to piglets was found to be materially beneficial in that, it increased growth rate of pigs and reduced mortality among them. The supplementary feeding reduced the demand on sow's milk and consequently the sow experiences less weight loss than when pasture feeding is not done. Piglets start nibbling at food when they are about a week old and by the time they are two to three weeks they will start consuming feed.

The milk production of the sow starts declining after third week of lactation. Therefore, it is necessary that the piglets are started on a solid feed by that time to meet their nutritional requirements. This is all the more necessary when the litter size is large and when the sows are in poor condition.

The pre-starter ration should be rich in protein and vitamins and low in fibre content. Adding a higher proportion of antibiotics has also been found to be beneficial; this ration usually has 24 per cent protein. This feeding starts usually when the piglets are one week old, weighing 2-2.2 kg. This extends usually to a period of two weeks during which each pig consumes one and half to two kg of pre-starter ration and gains about three kg, to weigh 5.0 kg at the end of the period. These rations are self-fed in a trough or fed in a pen separate from that of the mother adjacent to it. The gates may be adjusted so that the piglets can get in and eat but the sow cannot. They need plenty of fresh water.

Nutritional anaemia in piglets. The new born pig has only a limited reserve of iron for haemoglobin synthesis due to inefficient placental transfer and the transfer of iron into sow's milk is extremely inefficient. The newborn pig has only about 50 mg of iron store in the liver. But it daily needs about 5 to 10 mg of iron. Even with high dietary concentration of iron and copper, the milk of sow is deficient in these elements. These factors along with the high growth rate of piglets (3-6 weeks of age) causes a highly fatal disease of suckling piglets characterised by marked decrease in haemoglobin and degeneration of liver. The piglets become dull, inactive, fatigued and show lack of vigour and dyspnoea. This laboured breathing is called 'Thumps'. The piglets will have rough coat and they will be weak and thin. The skin over neck and mucous membrane become pale and piglets may develop diarrhoea. They die suddenly. The preventive methods are:

1. Add small quantity of ferrous sulphate (25 mg) and copper sulphate (5 mg) in the diet of pigs daily or these may be added in drinking water. Oral mineral tablets can also be used.
2. Paint the udder of the sow with the following mixture; ferrous sulphate- 50 g, copper sulphate 7 g, zinc sulphate 3 g and sugar/ jagary 50 g in one litre of water.
3. Intramuscular injection of iron dextran (Imferon®) 1ml on the 4th day and 14th day for each pig.

A wild pig does not develop nutritional anaemia because they have access to soil which is rich in iron. So this deficiency disease is restricted to farm reared pigs on concrete floors.

Fresh clean soil may be kept in the pen to supplement iron from soil but the pigs make mud puddle out of it and the chance for the transmission of parasitic ova are more. Therefore the safe and easiest method of combating piglet anaemia is to inject the piglets with 100-150 mg of iron in the form of iron dextran.

Feeding starter pigs

Starter ration otherwise known as creep feed, should contain 20-22 per cent protein. The rapid growth of the piglets with the concomitant increase in daily nutrient requirement, combined with the decline in milk yield, necessitates the provision of starter ration.

Creep feeding. The practice of self feeding easily digestible concentrate feed which contain 22 per cent crude protein and 3500 Kcal/kg DM of digestible energy to young piglets while still suckling the mother in a separate enclosure, which cannot be accessed by the sow is called 'creep feeding'. Creep is a device by which piglets are allowed access (creep in) to the concentrate mixture with the exclusion of sow. Creep feeding has pronounced effect on the livability as there is significant differences in mortality between those creep fed and not fed. It is fed to pigs between 5.0 to 15.0 kg live weights. Pelleting of feed is beneficial from this stage onwards. This period generally ends when the pigs are eight weeks old. They consume 9 to 11 kg of starter ration to gain about 10 kg. They should be provided at least 4 cm feeder space per pig. Fresh water should be provided *ad libitum*.

Feeding grower pigs

When the average weight of pig is 15 kg, the starter ration is replaced by grower ration, which has only 18 per cent protein. If weaned, the pigs are housed in small groups, less than 20 in number. Care should be-taken to see that pigs housed together are more or less of the same weight. Pigs consume 50-60 kg of grower ration to weigh about 35 kg at the end of this period.

It may be difficult to achieve these standards under feeding and management conditions available in the tropics. The period taken to reach various stages and the growth response under two levels of energy in the tropics are given in Table 3.1. The lower energy level is what is often obtainable in a tapioca based feed in tropics. The standards of feeding established in temperate regions and the growth responses are shown in Table 3.2 for comparison.

Growing and finishing pigs

This is the stage during which the weaned pigs grow from 35 kg to 60 kg body weight. Rapid and economical gains are possible if the pigs

TABLE 3.1 PERFORMANCE OF GROWING PIGS IN HOT-HUMIDTROPICS (KERALA) UNDER IDEAL MANAGEMENT CONDITIONS

Particulars	Pre-starter & Starter		Grower-1	Grower-11		Growing and Finishing	
Weight (kg)	2-10		10-20	20-35		35-60	
CP%	22		18	16		14	
	Ideal Maize based	Tapioca based	Ideal Maize based	Ideal Maize based	Tapioca based	Ideal Maize based	Tapioca based
DE, Kcal/Kg	3500	3500	3150	3300	2970	3300	2970
Days to attain	49	42	46	30	37	36	32
Feed consumed (kg)	Not Known	35	35	37	48	68	85
Daily gain (g)	163	238	217	500	405	700	600

Source: Kurian Thomas (1980) (Personal Communication).

TABLE 3.2 INTERNATIONAL STANDARDS OF FEED REQUIREMENTS & GROWTH EXPECTED IN PIGS

Period	Weight(kg)	Days	CP%	DE Kcal Per kg	Amount of feed Needed Per pig (kg)	Daily Gain Expected (g)
Pre-starter	2-5	1-2	24	3500	1.5-2.0	150
Starter	5-15	22-56	22/20	3500	9.0-11.0	300
Grower	15-35	57-90	18	3300	50-60	580
Growing & finishing	35-90	91-80	14	3300	140-150	600

have access to a good pasture and supplementary feeding is given. Vitamin B_{12} supplementation reduces protein requirement of weaning pigs, from 18 per cent to 14 per cent. With every 20 kg additional weight, a reduction of one per cent in protein is recommended.

Pigs above 60 kg live weight and until marketing are known as finishing pigs. The only major difference in the nutrient requirement of this group from the previous one is in their protein requirement. They need rations containing only 13 per cent protein. Under hot climatic conditions, it will be advantageous to market/slaughter pigs between 70-75 kg.

Feeding of Adult male pigs

A breeding boar requires 2-2.5 kg concentrate per 100 kg body weight depending on the age, condition and breeding demand. Feeding allowance should be so adjusted that the pig is neither fatty nor run down. Greens should be provided if kept indoors. The crude protein content of the boar ration should be 14%. Boars should be fed to keep them in thrifty condition avoiding fattening. The boar may be given almost the same kind of ration fed to the gilts during the breeding period. It is advisable to give one-quarter to one-half of an acre of good pasture to every boar to get exercise, as well as pasturage. They are usually given a supplementary ration by self-feeding method. If the boars get too fatty, the concentration of grains in the ration may be reduced and that of lucerne meal increased.

Feeding Adult Females

Feeding gilts: Gilts, like boars, should be fed limited amount of nutrients to keep them thrifty but not fat. If they are already fat, there will be no scope for flushing during the breeding season. The gilts meant for breeding should be removed from the growing–fattening group when they are between 45-57 kg body weight. They should be subjected to limited feeding at a range of about two-third of the requirement of growing-fattening group. It has been observed that limited feeding increases the litter size by one to two pigs over than when they are full-fed. It also reduced feeding costs.

Feeding un-bred sows: Un-bred sows may be fed similar to the gilts to keep them in medium condition. Usually, their requirements are less than that of gilts, the latter needing nutrients for growth. The actual amount or composition of the ration will depend on the condition of the sow. If the sow is lean and run-down in condition due to previous pregnancy and lactation, it needs more of nutrients to recoup.

Flushing. The feeding of gilts and sows more liberal ration two weeks before breeding until they are bred is called 'flushing'. This practice produced a gain of 0.7 to 0.9 kg per day and ensures good health. It improves the condition of gilts and sows there by increasing the number of ova shed, fertilisation and subsequent implantation of zygot. Flushing ration should havc plcnty of protcin, minerals, vitamins and green forages. Sows require less than 20 g of flushing ration for every kg body weight while gilts require 25 to 30 g as they are growing. The advantages of flushing are: It improves the physical condition of female pigs, increases ovulation rate, minimise embryonic loss, increases litter size, produce more uniform piglets, prompt post-weaning oestrus with prominent heat signs and shortens the period between weaning to successful conception.

Feeding pregnant sows: The ration influences to a great deal, the size and type of the litter born. Good feeding during gestation will result in a large litter, heavier and healthier pigs at birth, ample milk yield by sows, reduces the incidence of still births, abnormalities and runts. More and heavier pigs are weaned. Pregnant sow should get a ration which contains-

Protein	14 %
Calcium	0.7 %
Phosphorus	0.55 %
Vitamin A	5500 units/kg

Vitamin D	880 units/kg
Riboflavin	5.5 mg/kg
Niacin	33 mg/kg
Pantothenic acid	18 mg/kg
Vitamin B_{12}	22 pg/kg
Common salt	0.5 %

They may be fed on pasture with concentrate supplementation or on dry lot. The requirements may be met through grains, protein supplements, milk products and hay. The amount of feed to be fed to this group will depend on age, stage of pregnancy and the condition of the animal. Test conducted at the swine testing stations of Iowa, Illinois, Ohio and South Dakota indicated that 1.8-1.9 kg of feed per day will result in maximum birth weight and weaning weight of pigs. Sow should gain 27 to 36 kg during gestation period, compared to 32 to 41 kg by gilts. Lucerne pasture can provide most of the nutritional needs of the bred sows. But gilts need more supplementation.

Hand feeding or self-feeding technique can be adopted. In self-feeding, the proportion of the bulky feeds in the ration must be governed by the condition of the sows. Concentrates are increased if sows do not gain in weight properly. Usually, the ration of pregnant sows should have 25-35 per cent lucerne.

Feeding sow at the time of parturition and immediately after. As the time for farrowing approaches, the swine-farmer should exercise more and more vigilance and give particular care on their feeding. About four to five days before farrowing, the sow's ration should be reduced to nearly half. They should be prevented from becoming constipated by including laxative feeds like wheat bran, linseed meal, *etc.* in the ration. The amount of high energy feeds like maize is reduced to be substituted by feeds like wheat bran, lucerne hay or oats. On the day of farrowing, it is better not to give any feed at all. But enough of clean fresh water should be available throughout the period. The farrowing ration formulated at I.V.R.I, Izatnagar comprises of:

Maize	35 parts
Soya cake	8 parts
Wheat bran	35 parts
Fish meal	5 parts
Mineral mixture	1.5 parts
Salt	0.5 parts

It is essential not to feed the brood sow heavily after farrowing. If fed too much, she may produce excess milk over and above the needs of the piglets, which can lead to scours in them. The sow may be fed the same bulky ration as fed before farrowing for three to four days after farrowing. It is best to feed 50 per cent of the ration on the first day and to increase the ration by 200 to 300 grams daily until she is full fed. Another recommendation (Delaware Bulletin, 40) is to feed 400-500 grams of whcat bran and plenty of lukewarm water to the sow on the day after farrowing. It is also necessary to satisfy the appetite of the sow; otherwise, they may eat their own litter.

Feeding lactating sows. Sows receiving adequate ration produces 2.5 to 3.6 kg milk per day. This is adequate for small litters. But sows nursing more than ten piglets must produce to their maximum capacity to satisfy the nutritional need of the piglets. The weight as well as thrift and vigour of the weaning piglets is dependent to a great extent on the care and feeding of the lactating sow. The gain made by piglets in the first few weeks is directly related to the milk producing ability of the sow and the composition of milk of the sow. Sows must receive feeds containing the required quantity of nutrients sufficient to sustain lactation. The composition of sow's milk (Table 3.3) is the best guideline to estimate the required raw materials to be supplied for it.

A sow which produces about 180-185 kg milk during a lactation of 56 days, secretes in the milk about 12 kg proteins, 450 g of calcium and 280 g of phosphorus. A bulky ration may be continued up to the second week of lactation. Sometime during the second week, the sow should have started getting the full feed. By that time piglets requires large amounts of milk and the ration of the sow should be changed in such a way as to induce and sustain heavy lactation. Experiments conducted at I.V.R.I, Izatnagar showed that a lactating sow requires 3.5 kg concentrate feed plus 0.2 kg per piglets in the litter.

TABLE 3.3 COMPOSITION OF NORMAL MILK OF SOW

Water	8.75 %
Dry matter	19.25 %
Fat	6.5 %
Protein	6.5 %
Nitrogen Free Extract	4.9 %
Mineral Matter	0.96 %
Calcium	0.25%
Phosphorus	0.16%
Vit. A, I.U per g of fat	18.50
Vit. C, mg/100ml	12.20
Thiamine, micrograms/100ml	67.70
Ribofalvin, micrograms/100ml	45.70

Source: Bowland *et al.* (1986)

The dry-lot lactation ration should contain 15 per cent crude protein for both first litter gilts and sows. If on the other hand, they are put on a legume pasture, 13 per cent crude protein will be sufficient in the ration. Lactation rations should contain 0.7 per cent calcium, 0.6 per cent phosphorous and 0.5 per cent common salt. They require vitamins in the following amounts under dry-lot feeding: Vitamin A -1500 I.U. Vitamin D -100 I.U. Riboflavin- 1.5 mg, Niacin-8.0 mg, Pantothenic acid 6.0 mg, Vitamin B_{12}-5.0 micrograms. Lactating sow requires 14 to 23 litters of water per head per day. The exact quantity consumed will be determined by the weight of the pig, its condition, milking capacity, climate and the litter size.

Self-feeding has been proven to be a very satisfactory method for lactating sows. By the judicious use of lucerne meal, wheat bran and oats, it is entirely feasible to self-feed sows in the early part of lactation when they should be given limited feeding. The ration should then be progressively made more concentrated as the nutrient requirement of the pig (through milk) increases. However, in India where labour charges are cheaper, there is very little to gain from self-feeding. Hand feeding may still remain a more economic proposition.

Feed additives

Substance not classified as nutrients, but added to the feed in micro quantities to improve the animal productivity, is known as feed additives. These substance can be defined as ingredients or combination of ingredients added to the diet of the animal to fulfil a specific need and/ or to boost up the productivity.

Feeding thyroprotein: Iodinated casein included in the ration of the sows has increased the gain of pigs by 20-30 per cent during the first week of lactation. Iodinated casein has one per cent thyroxine activity. It increases the metabolic rate and milk yield. The enhanced milk production results in heavier litter at the time of weaning.

Feeding antibiotics: Antibiotics are used as feed additives to stimulate growth, improve feed efficiency, secure uniformity of growth and control infections. Antibiotics increase feed efficiency of the growing and finishing pigs up to five per cent, reduces scouring and unthriftiness, increases rate of gain, improve feed conversion efficiency and reduces runts, leading to a uniform crop of pigs. A combination of broad spectrum and narrow spectrum antibiotics gives better result than either of these. Eg. Aureomycin and Terramycin. The recommended dose of antibiotics (aureomycin and terramycin) is 10-50 g/tonne of feed. The beneficial action of antibiotics as feed additives may be due to:

i) **Sparing action of antibiotics:** Antibiotics spare nutrients by favouring either more efficient utilisation of these or by their increased synthesis in the digestive tract.

ii) **Thinning of mucosa:** The beneficial effects of antibiotics may be due to enhanced absorption caused by the thinning of intestinal mucosa.

iii) **Action on micro-organisms:** Antibiotics promote the growth and increase in the number of certain micro-organisms which produce known or unknown growth stimulating substances. Antibiotics decrease the number of micro-organisms which use up growth factors in the intestinal tract.

Antibiotics decrease the number of micro-organisms which are pathogenic or produce toxins.

The type of antibiotics should be changed annually to reduce the likely hood of built-up resistance to a particular antibiotics within a herd.

Adding copper sulphate in combination with antibiotics improve performance more than using either alone. But due to the emergence of Anti Microbial Resistance (AMR), feeding of antibiotics as feed additives has been discouraged by law.

Soft Pork

In swine, the feed fats are deposited in the body without undergoing much change. Thus when finishing pigs are liberally fed with feeds which contain large quantities of short chain unsaturated fatty acids like rice bran, groundnut, cake, vegetable oil etc, it results in soft fat in pork. Soft pork is undesirable because, it is slimy to touch and unattractive. The meat is difficult to handle or cut since oil drips from it. There will be more shrinkage when smoked and more wastage during storage and cooking. Therefore, finishing pigs should only be fed with ration producing hard fat. It has been found that when rations producing hard fat is given following a period of feeds rich in unsaturated fats, the body fat gradually becomes harder. It has also been found that this process take place more rapidly if the animals are first fasted for a period before the change in ration is made. This practice is called 'hardening off'.

Deficiency Disease of pigs

Pigs being omnivorous, monogastric, simple stomached animals lack the benefit of microbial digestion such as, manufacture of 'B' vitamins and utilisation of roughages. Therefore, feeding of swine is very crucial and if not on scientific grounds, leads to deficiency diseases. The symptoms of deficiency diseases in pigs along with the corresponding nutrient is given below:

Nutrient	Deficiency Symptoms
Vitamin A	Reduced weight gain, in-coordination, posterior paralysis (degeneration of sciatic and femoral nerves), blindness, increased cerebrospinal fluid pressure, atrophy of epithelial layer of genital tract, reduced fertility, resorption of foetuses.
Vitamin D	Rickets, Osteomalacia, Eclampsia, enlargement of joints, lameness, tetany, rib and vertebrae fracture.
Vitamin E-Selenium	Mulberry Heart Disease (Cardiac muscle degeneration), Hepatosis dietetica (Liver necrosis), White muscle disease, degenerative thrombotic vessel injury.
Vitamin K	New born pigs with loss of blood from umbilical cord, Internal haemorrhage, sudden death.
Biotin	Excessive hair loss, skin ulceration and dermatitis, exudates around the eyes, transverse cracking of hooves, spasticity of hind legs, scaly skin.
Choline	Lowered reproduction, spradled legs, rough hair coat, fatty infiltration of liver and kidney.
Folic Acid	Slow weight gain, fading hair colour, bone marrow hyperplasia, normocytic anaemia.
Niacin	Necrotic leasion of intestine, Ulcerative gastritis, severe diarrhoea, rough hair coat, hair loss.
Pantothenic Acid	Abnormal movement of the hind legs called "Goose stepping", slow growth, diarrhoea, anorexia.
Riboflavin	Slow growth, seborrhoea, cataract, stiff gait, alopecia, poor conception.
Thiamine	Loss of appetite, reduced weight gain, myocardial degeneration, sudden death, cardiac hypertrophy.

Pyridoxine	Reduced appetite and growth, exudate development around the eyes, convulsions, ataxia, epileptic seizures, coma and death, microcytic hypochromic anaemia, fatty infiltration of liver.
Vitamin B_{12}	Reduced weight gain, loss of appetite, irritability, hypersensitivity, rough skin and hair coat, hind leg in-coordination.
Vitamin C	Petechial haemorrhages throughout the body.
Calcium and Phosphorus	Depressed growth and poor bone mineralisation resulting in rickets in young pigs and osteomalcia in older swine, beading of ribs, lameness and tetany, posterior paralysis.
Sodium and Chlorine	Deprived appetite, unthrifty condition.
Magnesium	Hyper irritability, muscular twitching, reluctance to stand, weak pastern. loss of equilibrium, stepping syndrome, tetany.
Potassium	Anorexia, rough hair coat, emaciation, ataxia, cardiac impairment.
Copper	Poor iron mobilisation, abnormal haemopoisis, poor keratinisation and synthesis of collagen, elastin and myelin, microcytic hypochromic anaemia, bowing of legs, spontaneous fracture, cardiac and vascular disorders.
Iodine	Stunted growth, lethargic, elongated thyroid, weak or dead piglets that are hair less, Myxoedema.
Iron	Anaemia, poor growth, listlessness, rough hair coat, wrinkled skin, pale mucous membrane, laboured breathing with a spasmodic jerking of the diaphragmatic muscle called "Thumps". Hypochromic microcytic anaemia, enlarged fatty liver, ascites.

Manganese	Abnormal skeletal growth, increased fat deposition, irregular or absent oestrus cycle, reabsorbed foetuses, small weak pigs at birth, reduced milk production.
Selenium	Sudden death is a prominent feature of selenium deficiency. Hepatosis dietetica, white muscle disease, mulberry heart disease. Impaired reproduction, reduced milk production and impaired immune response.
Zinc	Poor growth, poor appetite, parakeratosis and retarded testicular development.

CHAPTER 4

SWINE HOUSING

Housing of pigs

The basic objectives for housing pigs are:

1. Provision of health, sustaining comfortable environment to the pigs and piglets
2. Provision of desirable working condition for those working on the farm.
3. Integration of housing with feeding, watering and manure handling systems.

   Sheds for pigs are known as sties (singular sty). There shall be sties to house pigs of different ages separately.

Building Systems

Three types of building systems can be employed for successful piggery rearing operations.

All in one/birth to market building system: In this system, there will be only one centrally located building and the pigs remains in the same building during all the phase of their life-cycle. The building cost is low and this system requires least labour. The microenvironment of the building can be controlled under extreme weather condition. The major drawback of this system is the increased incidence of disease.

Two-building plan: In this system, there are only two types of buildings *i.e.* the farrowing shed and the growing and finishing shed. The piglets from birth to weaning *i.e.* up to 11.4 kg remains in one building and the pigs from weaning to marketing i.e. from 11.4 to 91 kg remains in another building. The advantages and disadvantages are midway between the all-in one system and the three building system.

Three building system: In this system there are 3 types of buildings, *viz.* farrowing sheds, growing and finishing sheds. The main advantage of this system is that, incidence of disease is less and efficient isolation is possible. The disadvantages are the construction cost is high and labour requirement is maximum. The movement from one building to another also causes stress to the pigs.

Location of Buildings

The specification for the location of the farm buildings are as given below:

1. The structure should be on a fairly raised and well drained site not liable to flooding.
2. The structure should be at least 45 m away from dwellings and factories. 30 m away from dairies, poultry houses and food grain storage structures. 45 m away from a fire source and 1 km away from garbage dumping ground, slaughter houses, hide curing centres and tanneries.
3. The boundaries of the site should be at least 50 m away from the nearest transit road and should be properly fenced. Suitable spaces should be left for sanitary purposes when locating building and structures.
4. Adequate supply of water must be ensured at the selected site.
5. The structures should be located and oriented as to avoid heavy draught but at the same time, have plenty of sunlight and circulation of fresh air.

Layout

The layout of the buildings has marked influence on the work efficiency. Most important is the facility for proper sanitation with a good drainage system. Other factors are:

1. Drainage should be away from farm houses and other buildings. The pasture drainage should not enter waterways leading to streams or lakes.
2. Space for more buildings should be left in respect to feed store, other store and roads.
3. Manure disposal facilities should be located downwards from houses to minimise odour problems.
4. Facilities for movement of animals, operations, equipments *etc.* should be planned in the layout.

General Housing Guidelines

1. The house should give adequate protection against direct sunlight and rain as pigs are very sensitive to heat and humidity.
2. The floor and wall should be strong to with stand the rooting behaviour of pigs. Concrete flooring is durable and easy to clean. The walls may be of bricks, finished smoothly and doors of strong wooden planks or iron.
3. Feed troughs should be placed along the front to facilitate feeding from outside.
4. Pigs being non sweating animals, thrive well in the temperature range of 20-25 °C. Provide shades and cooling devices such as sprinklers, foggers *etc.* to maintain thermal comfort.
5. Design should be such that all animals are observable easily from outside and the labour requirement is less.
6. Boars, pregnant and dry sows and growing pigs are usually kept in open yards with partially sheltered areas. Farrowing sows are housed in completely enclosed houses or pens.

7. Simple low cost houses constructed with locally available materials as per the above guidelines are preferred in rural areas. Multipurpose pens which can be used for all categories of pigs can also be designed meeting the floor space requirement (Table 4.2).
8. Individual or group housing in cages made up of vertical G.I pipes/ M. S rods and also farrowing crates can be adopted in large high-tech farms (Table 4.3).
9. Un-castrated male and females should not be housed together beyond the age of 4 months.

Environmental requirements of pigs

Temperature: The critical temperature of new born pig is 30 °C which gradually declines as the pigs grow older to 20 °C. The normal deep body temperature of adult pig is 39 °C. New born pigs are likely to be affected by chills and suffer from hypoglycaemia; therefore supplemental heat have to be provided during the first few days of post natal life, but adults pigs are poor thermo regulators and require cooling devices. When temperature is low, the pigs huddle together and the lying area become dirtier. When temperature increases, the pigs lay apart and the likely hood of tail biting increases. The temperature requirements for different categories of pigs are given below.

TABLE 4.1 COMFORT ZONE FOR DIFFERENT CATEGORIES OF PIGS

Body weight	Temperature range
New born piglets	30-35 °C
Nursing piglets < 10 kg	26-30 °C
Weaned pigs- 10-15 kg	22-26 °C
Growing pigs - 15-30 kg	18-22 °C
Adult pigs - 30-60 kg	16-20 °C
Finishing pigs > 60 kg	14-20 °C
Gestating sow	16-20 °C
Boars	16-20 °C

Humidity: Relative humidity up to 85% has not much influence on pig's performance. But when above 85% along with high temperature is detrimental for pigs. Under this stress, the pigs resort to panting. The condition in the sties should neither be dry and dusty to cause respiratory distress, nor so damp as to cause condensation. For optimum performance, the humidity should be between 60-80%.

Ventilation: Adequate ventilation is important in pig houses to prevent built up noxious gasses, dust and unpleasant odours. In buildings with slotted floors, the excreta remains for more time leading to production of gases like carbon dioxide, ammonia and hydrogen sulphide, which are dangerous to animals and man. There should be slight air movement at the level of animals. An air speed of 0.1m/s at pig height will offer good ventilation.

Acceptable levels of gasses in the environment.

Carbon monoxide	5 ppm
Carbon dioxide	2000 ppm
Hydrogen sulphide	0.1 ppm
Ammonia	10 ppm

Levels above these are indicative of inadequate ventilation.

The main systems of ventilation are:

1. Natural cross and roof ventilation. Natural air flows inside and outside of the houses through windows place opposite to each other.
2. Forced draft ventilation. In which fresh air is forced into the building and stagnant air allowed to flow out using fans.
3. Exhaust ventilation. In which stagnant air is forced out by exhaust fans and fresh air drawn in.

Floor Space Requirements for Pigs

TABLE 4.2 FLOOR SPACE REQUIREMENTS FOR PIGS

Type of animal	Floor space requirement (Sq. M per animal)		Maximum number of animals per pen
	Covered area	Open paddock	
Boar	6.0-7.0	8.8-12.0	Individual pens
Farrowing pen	7.0-9.0	8.8-12.0	Individual pens
Fattener (3-5 months old)	0.9-1.2	0.9-1.2	30
Fattener (above five months)	1.3-1.8	1.3.1.8	30
Dry sow/gilt	1.8-2.7	1.4-1.8	3-10

Houses for Different Categories of Pigs

Boar Sty: The boar should be housed in individual stalls. There should not be more than 24 pens under one roof. Acute heat stress (exposure to environmental temperatures of 33.3 °C for 72 hours) reduces total number of sperms per ejaculate, reduces sperm motility, increases abnormal sperm per ejaculate and reduces fertility. Therefore cooling device like sprinkling, fogging or wallows is mandatory in the breeding boar pen (Fig. 4.1).

FIG. 4.1 FOGGER ALONG WITH CEILING FAN

Dry sow and gilt sty: Dry sow and gilts can be housed together in small farms, but in large farms, there shall be different sties for dry sow and gilts. These are simple loose houses in which, the floor shall be cement paved in the covered area and brick-lined in the open area. There should not be more than 40 pens under one roof and each pen should accommodate not more than 5-10 pigs depending on the stage of pregnancy.

Farrowing-cum-nursing sty: Totally enclosed sties with no large openings to the outside are provided as farrowing and nursing sties for providing insulation. There should not be more than 40 pens under one roof and each pen may accommodate one sow with its litter (Fig. 4.2). Adult females in advanced pregnancy should be transferred to this pen two weeks before the expected date of farrowing and left there during the nursing period. To reduce overlaying mortality and for better management, the farrowing sty should be fitted with farrowing crate or guard rails. Provision for creep feeding and infrared lamp to provide warmth to the piglets should also be provided.

FIG.4.2 FARROWING CUM NURSING STY

Farrowing crate (Fig. 4.3) is a tubular crate made of G.I pipes, which will prevent the sow flopping from the standing position directly on to her sides thus considerably reducing overlaying mortality. The structure forces the sow to lie down on her belly first before rolling over on to either side, at the same time provide enough space for the sow to stand up and lie down but restricting lateral movements. It also protects the operator as some sows become aggressive after farrowing and attacks the operator. The farrowing crate has the following dimension; total width 1.4-1.5 m, length 2-2.4 m. 0.6 m width is needed for the sow and about 0.4 m is needed on each side for the baby pigs. The initial investment for this facility will be high, but later will be compensated by more number of pigs produced. New animal welfare laws discourage the use of farrowing crates.

Guard rails are 5 cm diameter G.I pipes, fixed 25 cm above the ground level and 20 cm from the wall, provided all along the walls (Fig. 4.4). Piglings can run under these rails when the sow lies down so that they may not get crushed between the sow and the wall.

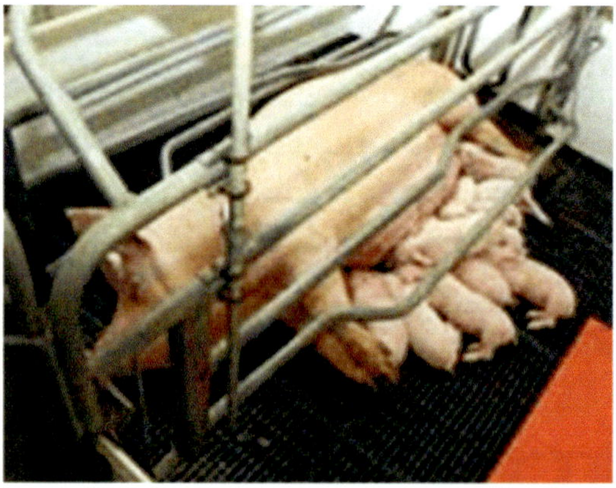

FIG 4.3 FARROWING CRATE ARRANGEMENT

FIG 4.4 GUARD RAILS IN FARROWING STY

Fattening sty: These are for housing pigs which are being fattened for the market. Each fattening sty should not have more than 30 pens under one roof, each accommodating 15-30 pigs depending on their age.

Weaner pig sty: Piglets that are weaned are housed in these pens up to six months of age. Each sty should not have more than 30 pens under one roof, each accommodating not more than 30 piglets in each pen.

Sick Animal Sty: An accommodation of 5 per cent of the stock should be provided for nursing sick pigs. This sty should be at a reasonable distance from other sites for efficient isolation. Arrangement for feeding and watering should be made in the sty itself.

Accessory Items

Feeding troughs: Fixed concrete troughs with smooth finishing are usually provided. The height of feeding trough from floor to pig side should be 15 cm while it should be 38 cm wide and divide at an interval of 30 cm to prevent pigs from lying inside the trough by placing iron bars across it. Now-a-days stainless steel feeding trough are fabricated which can be easily cleaned and is durable.

Nipple drinkers: The main advantage of nipple drinker is that they reduce the contamination of drinking water and reduce the incidence of various diseases (Fig. 4.5). Pigs can drink water as per their requirement at any part of the day or night (Fig. 4.6). The wastage of water is also reduced considerably. Pigs learn fast to drink from nipple. One nipple per 15 growing pigs should be provided. It also saves labour.

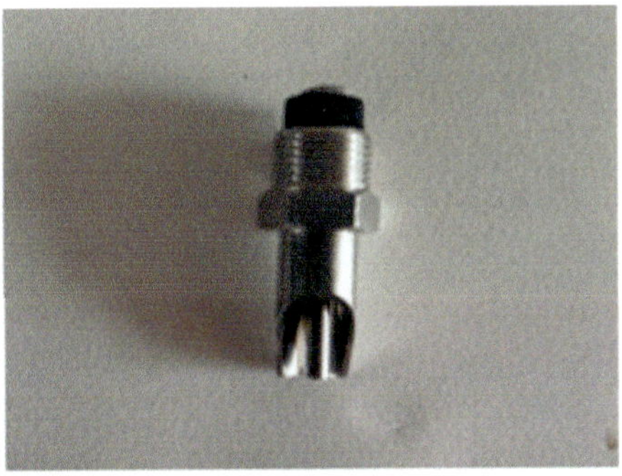

FIG. 4.5 PIG DRINKING NIPPLE

FIG. 4.6 PIGS DRINKING FROM NIPPLE DRINKER

Hog Wallows: It is a water pool where hogs enter into water and cool their body in summer. Breeding and pregnant animals should be provided with wallows. Fattening pigs show habitual wallowing and take more feed to maintain body temperature, making them unthrifty. Therefore, it is better not to provide wallowing tanks for growing and fattening pigs. A masonry wallow made of cement concrete with proper drainage system is to be made. A wallow of 3 metre length, 2 metre width and 45 cm depth will easily hold a herd of 20 hogs. Top side and end walls of wallow should extend at least 10 cm above the surface of the floor to prevent surface water running into wallows. The entrance and exit incline must have a gradual and easy slope so that hog may enter and leave wallows conveniently.

Water sprinkling and internal fogging are the best methods to reduce heat stress in pigs. When water sprinklers are used, one nozzle per 25 to 30 pigs is sufficient (Fig. 4.1).

Manure pit: Manure pits shall be constructed at a distance of not less than 15m away from the sties. Manure pits can be one or two depending on the stock of pigs. Liquid manure, including urine and washings can also be disposed for irrigation of cultivated areas.

Bedding: 10-15 cm of chopped straw should be provided as bedding for the first 3 days after farrowing to give warmth, softness and non slippery floor. The bedding must be kept clean dry and evenly distributed.

Apart from the building for the stock, a pig farm should have an office room, a feed grinding and mixing room, feed store and facilities like weighing yards, loading and unloading ramps, store for keeping miscellaneous items and garbage boiling vats.

There should be a simple weighing yard and sorting pen for the pigs. This is the place where pigs are periodically collected and weighed to assess their growth and weight gain. There shall be a permanent ramp in swine farm for loading and unloading pigs into and out of the truck. There shall also be a boiler room in which provisions are made for boiling garbage (before feeding to pigs) in big vessels or in jacketed vats with steam generated in the boiler.

Details of Construction

Floors: The floor should be hard, impervious smooth, non slippery and easy to clean. Provide enough slope from one end to other for easy drainage (about 3 cm for every 2 m). Three types of floors are used in swine houses: Solid, partially slotted, and totally slotted.

Solid floors are usually made of concrete. The major disadvantage of solid floors is the difficulty in handling manure. Floors should be cleaned regularly which require more labour.

Partially slotted floors are a combination of solid and slotted floor. Some hand cleaning of manure is required with partially slotted floors.

Totally slotted floors almost eliminate the handling of manure. However, the cost of slotted floors is more than that of solid floors. Slotted floors are harder on the feet and legs of hogs. It is harder to control the temperature in a slotted floor building.

Commercial slats are available in concrete, metal, fibreglass and plastic. Metals used include steel, stainless steel, and aluminium. Though metal slats may corrode over long periods of time, concrete slats generally lasts the life time of the building. For baby pigs in farrowing pen, narrow slats, with spacing of 0.9 cm should be used. Wider slats, with spacing of 1.9 to 2.54 cm may be used in finishing houses. In the farrowing pen, the slats should be placed parallel to the sow for better footing. Manure is handled as liquid when slotted floors are used.

Walls: Walls must be strong and smooth. The height of solid brick wall can be between 0.93 to 1.22 m. Two Iron tubular pipes of 7.5 cm diameter are provided as railings at distance of 30 cm from each other.

Roof: The roof should be water proof and materials should be bad conductor of heat. Roofing materials can be asbestos sheets, G.I or aluminium-zinc corrugated sheets, tiles or coconut leaves. If G.I corrugated sheets are used insulation should be provided below it to reduce heat.

Doors: The doors should be fitted close to the floor so that pigs are not able to lift them by putting their snout underneath. Doors may be prepared by metal rods. The width of the door may be 75-90 cm and height 90 cm.

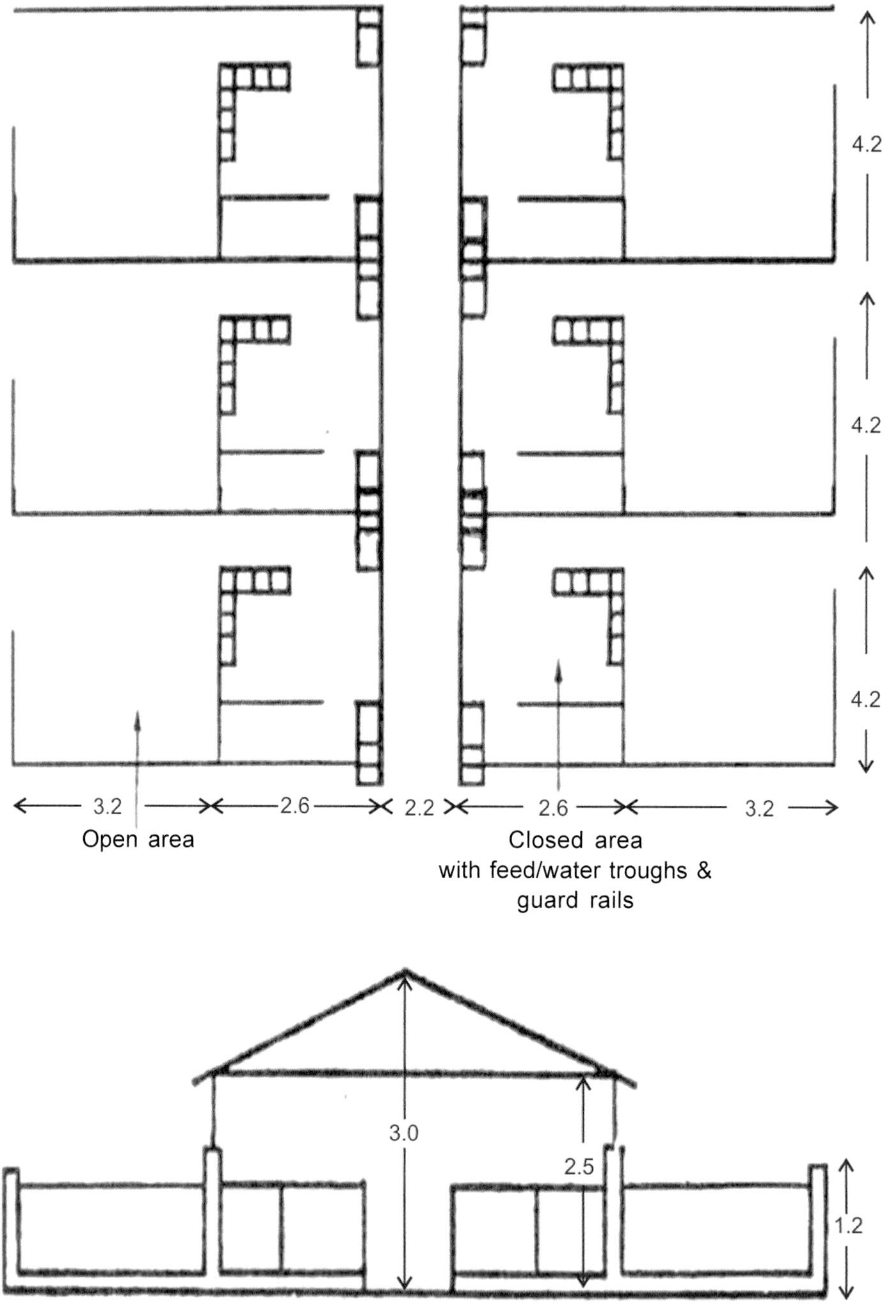

FIG 4.7 PLAN AND SECTIONAL VIEW OF A GROWER STY

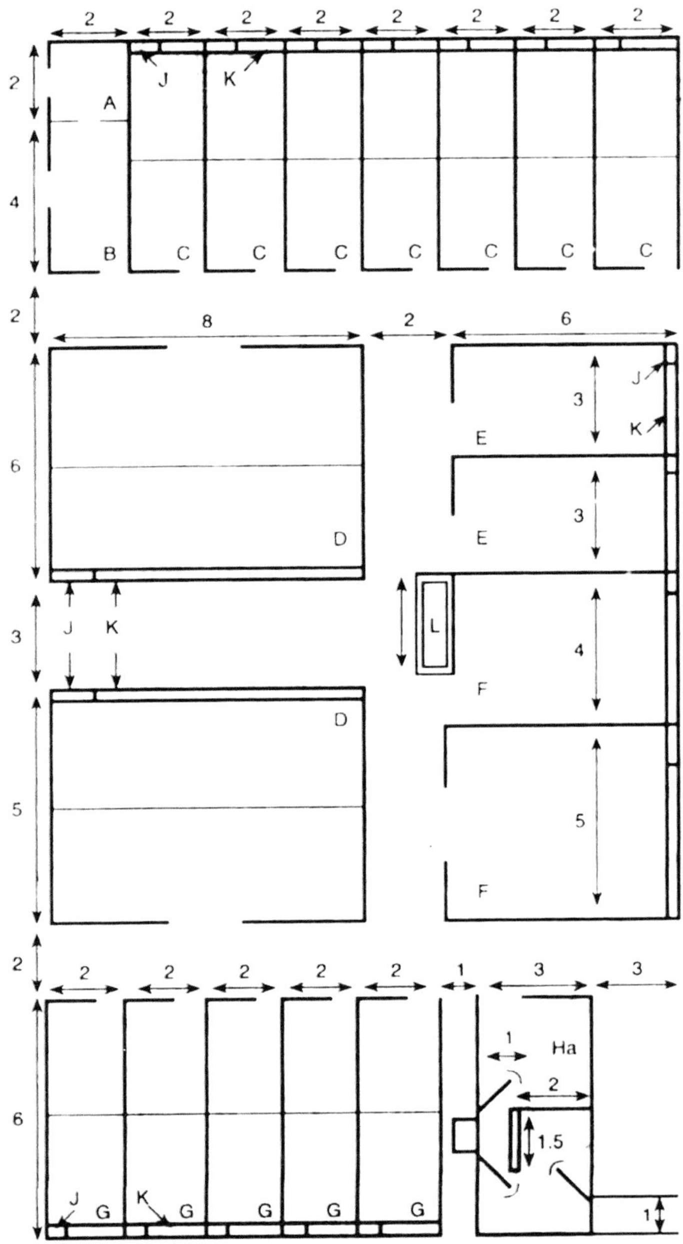

FIG. 4.8. LAY OUT OF BUILDING REQUIRED FOR A 50-SOW SWINE FARM

A. Office, B. Store, C. Farrowing pen, D. Dry sow/gilt, E. Weaner pig sties, F. Fattening pig sties, G. Boar sties, H. Weighing yard- a) Assembly pen, b) Weighing machine, c) Weighing platform, d) Collecting pen, I. Loading –unloading ramp, J. Water tank, K. Manger, L. Garbage boiling vat.

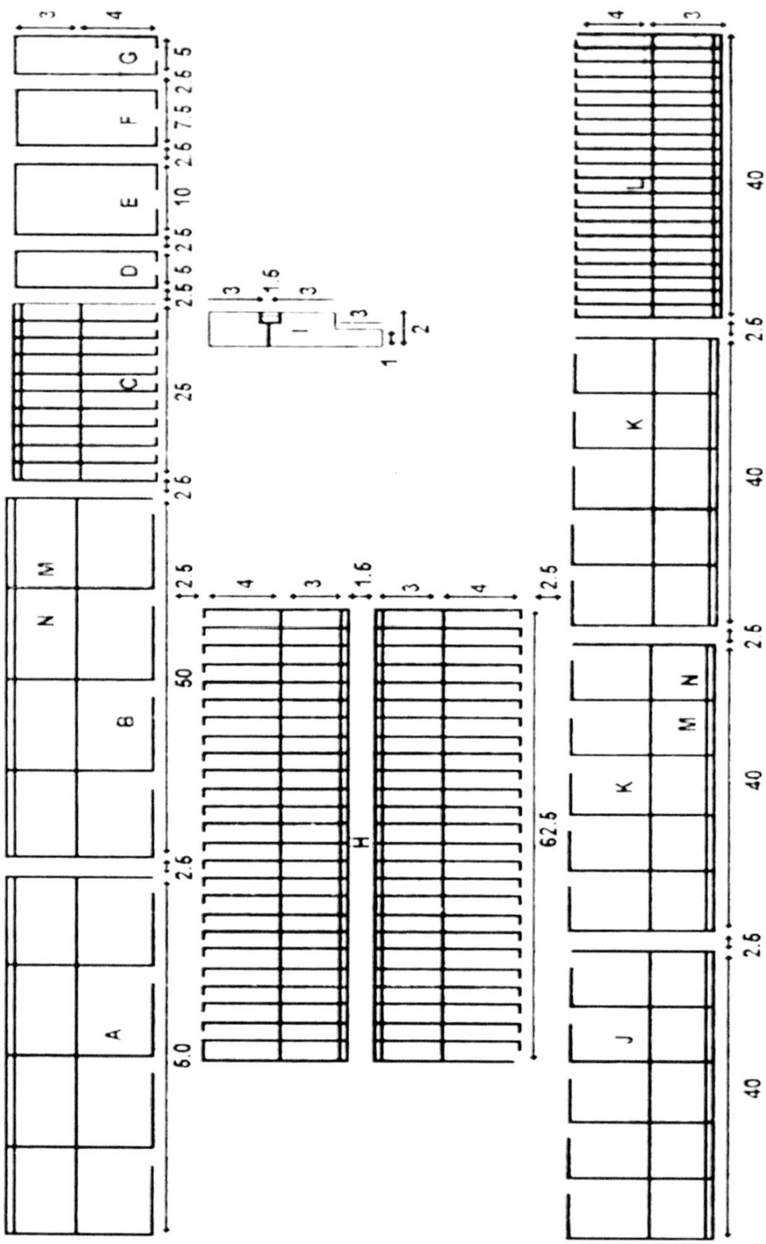

FIG.4.9 LAYOUT OF BUILDINGS REQUIRED FOR A LARGE (300 SO WS) SWINE FARM. A. Weaner pig sty, B. Fattening sty, C. Sick pig sty, D. Garbage boiling room, E. Slaughter room, F. Store, G. Office-cum-record room, H. Farrowing sties, I. Weighing yard with attached ramp, J. Gilt sty, K. Dry sow sty, L. Boar sties, M. Water tank-should be of 1/10 length as that of manger in each sty, N. Manger

Manure Handling

The excreta from the pigs should be disposed off in such a way that odour, flies and other nuisance do not develop. A 50 kg pig produces 4.38 kg manure daily which contains 140 g protein, 7 g phosphorus and 16 g potassium. Attempts have been made to utilise this nutrients contents by integrating pig, duck and fish mixed farming system, the algae and weeds of the pond utilises the nutrients of the manure for growth and proliferation, which are utilised by the fish and ducks.

Manure can be handled as solid or liquid. Solid manure is loaded on to a wheel barrow using shovel and spread into a pit for decomposition. Such manure will return 75% of its fertilising value to the soil. The volumetric capacity of fresh manure is 700 to 900 kg/m^3. Liquid manure is stored in a separate storage tank or in a pit under a slotted floor. The tank must be located at least 30 m from the water supply. Pumps and agitators are required to stir and remove the liquid manure and it is disposed off in a tank or wagon for irrigating agricultural land (Fig. 4.10).

Fig. 4.10 MANURE SPREADING USING WAGON

Dung de-watering Machine

Dung de-watering machine is a new addition in manure management. It is a de-watering screw press machine, which reduces the moisture content of the dung to 20% (Fig. 4.11). The dung mixed with equal quantity of water is pumped up by a submersible 2 HP chopper pump to the screw press, where the screw conveyor transports the material towards the outlet of the separator and it is pushed by the back pressure disc, to squeeze out the filtrate. The solid phase is separated from the liquid phase, which flows through the screen. The solid fraction is pushed towards the outlet on which is installed a counter-pressure diaphragm leading to the formation of a material plug. The dried dung is discharged at the end of the screw in a powdered form. The separated water goes through the screen, which flows into the collection box from where it is pumped for irrigation. This is a good machine to convert animal excreta to odourless organic manure for cultivation. The throughput is 3 to 4 tonnes per hour. The power source is a 3 HP motor for the screw press. A part of the de-watered liquid is again pushed back to the manure collection pit which aids in automatic agitation and prevents settling of large solid particles in the manure collection pit. The manure is rich in organic carbon, phosphorus, potassium, calcium and magnesium. The de-watered liquid has phosphorus and potassium more than the solid fraction which can be used for fodder cultivation.

FIG. 4.11 DUNG DE-WATERING MACHINE

Composting

Composting is the natural process of 'rotting' or decomposition of organic matter by microorganism, under controlled conditions. Animal excreta especially pig excreta can be used as a raw material for composting. The process of composting includes four main phases, which are 1) the initial phase, 2) the thermophilic phase, 3) the mesophilic phase, and 4) the maturation phase after which the compost can be used as organic manure. The degree of organic matter humification is generally accepted as a criterion of maturity or stabilization of compost. The humification process produces functional groups, and so increased oxidation of the organic matter leads to rise in cation exchange capacity. So, compost with high cation exchange capacity is regarded as an index of maturity. The degree of maturity can also be revealed by biological methods involving seed germination and root length, since immature composts may contain phytotoxic substances such as phenolic acids and volatile fatty acids.

Types of composting

Composting may be divided into two categories by the nature of the decomposition process. 1) Anaerobic composting 2) Aerobic composting. In anaerobic composting, decomposition occurs where oxygen is absent or in limited supply. Under this method, anaerobic micro-organism dominate and develop intermediate compounds including methane, organic acids, hydrogen sulphide and other substances. In the absence of oxygen, these compounds accumulate and are not metabolised further. Many of these compounds have strong odours and some present phytotoxicity. As anaerobic composting is a low-temperature process, it leaves weed, seeds and pathogens intact. Moreover, the process usually takes longer than aerobic composting. These drawbacks often offset the merits of this process.

Aerobic composting takes place in the presence of ample oxygen. In this process, aerobic microorganisms break down organic matter and produce carbon dioxide, ammonia, water, heat and humus, the relatively stable organic end product. Although aerobic composting may produce intermediate compounds such as organic acids, aerobic micro-organisms decompose them further. The resultant compost has little risk of phytotoxicity. The heat generated accelerates the breakdown of proteins, fats and complex carbohydrates such as cellulose and hemi-cellulose. Hence, the processing time is shorter. Moreover, this process destroys many microorganisms that are human or plant pathogens as well as weeds. It is considered more efficient and useful than anaerobic composting for agricultural production.

The aerobic composting process consists of

1. Properly mixing the waste with carbon rich material (straw, wood shavings, coir pith) in aerobic bins (Fig. 4.12). Carbon to nitrogen ratio of 20-25: 1 is usually recommended. Pure manure can also be composted if all factors are carefully monitored.
2. Addition of air by periodic stirring
3. Proper balancing of moisture levels (35 to 50% moisture)
4. Temperature monitoring to determine if composting conditions have occurred. The optimum temperature for composting is 54 to 66 °C. If the temperature fall below 49 or goes above 83 the compost pile should be aerated otherwise it will result in poor compost

FIG. 4.12 AEROBIC COMPOSTING
Source: Dr. K. Sivakumar, VCRI, Namakkal

5. True composting require 45-90 days to process and when finished, the material is stable with no further temperature build up when used or stored.

To facilitate the composting process and to prevent odour, the pit must be water tight so that outside water cannot enter. When managed correctly, there is little or no odour and there is no problem with flies.

Vermin Composting

It is the end-product of the breakdown of organic matter by some species of of earth worm (Udrils) (Fig. 4.13).

Benefits of vermin composting

1. 10-12% increase in production for any food crop, when compared with fertilizers
2. High nutritive value than compost
3. Have growth promoting substance like Indole-3-acetic acid
4. Seeds of weeds, toxins and other pollutants get destroyed when it passes through the earthworm gut
5. Fetches high price as it is used for 'organic farming'

FIG. 4.13 VERMIN COMPOSTING

Bio-gas Production

Bio-gas has assumed considerable importance as an alternative to conventional cooking gas throughout the world. Bio-gas production is a complex biochemical reaction carried out by the anaerobic bacteria in four steps *viz.* hydrolysis, acidogenesis, acetogenesis and methanogenesis. Bio-gas consists of the following mixture of gases.

TABLE 4.3 COMPOSITION OF BIO-GAS

Gases	Percentage
Methane (CH_4)	50-70%
Carbon dioxide (CO_2)	30-40%
Hydrogen (H_2)	5-10%
Nitrogen (N_2)	1-2%
Water vapour (H_2O)	0.3%
Hydrogen Sulphide (H_2S)	Traces

Certain bacteria which grow anaerobically on cellulosic material produce large amount of methane along with CO_2 and H_2O. These bacteria are collectively called Methanogens. Three major groups of bacteria hydrolytic, acidogens/acetogens and methanoges are responsible for breaking down the complex polymer in biomass waste to form bio-gas at anaerobic condition.

Hydrolysis

Hydrolysis is theoretically the first step in anaerobic digestion, during which the complex organic matter (polymers) are decomposed into smaller units.

Acidogenesis

During acidogenesis, the products of hydrolysis are converted by acidogenic (fermentative) bacteria into methanogeneic substrate.

Acetogenesis

Products from acidogenesis, which cannot be directly converted to methane by methanogenic bacteria, are converted into methanogenic substrates during acetogenesis.

Methanogenesis

Methanogenesis is a critical step in the entire anaerobic digestion process, as it is the slowest biochemical reaction of the process. 70% of methane originate from acetate while the remaining 30% is produced from conversion of hydrogen (H) and carbon dioxide (CO2).

Parameters affecting bio-gas Production

The parameters which affect bio-gas production are:

Feed stock characteristics, pH (hydrogen ion concentration), Hydraulic Retention Time (HRT) is the time taken by the substrate for maximum gas production and Carbon Nitrogen ratio (C: N ratio). The optimum levels of bio-gas parameters are given below.

TABLE 4.4 OPTIMUM LEVELS OF BIO-GAS PARAMETERS

Parameters	Optimum level
Digester Temperature	32-40 °C
Digester pH	6.8-7.2
Carbon Nitrogen ratio	30 :1
Total solid content of substrate	7-10%
Hydraulic Retention Time	30-50 days

Basic models of bio-gas plants

The basic bio-gas plant models are the floating drum model and the fixed dome model, which are detailed in Fig.4.14

Bio-gas production from different feed stock of animal origin

The different categories of animal feed stocks, bio-gas yield per kg and average bio-gas yield per animal is furnished in Table 4.5

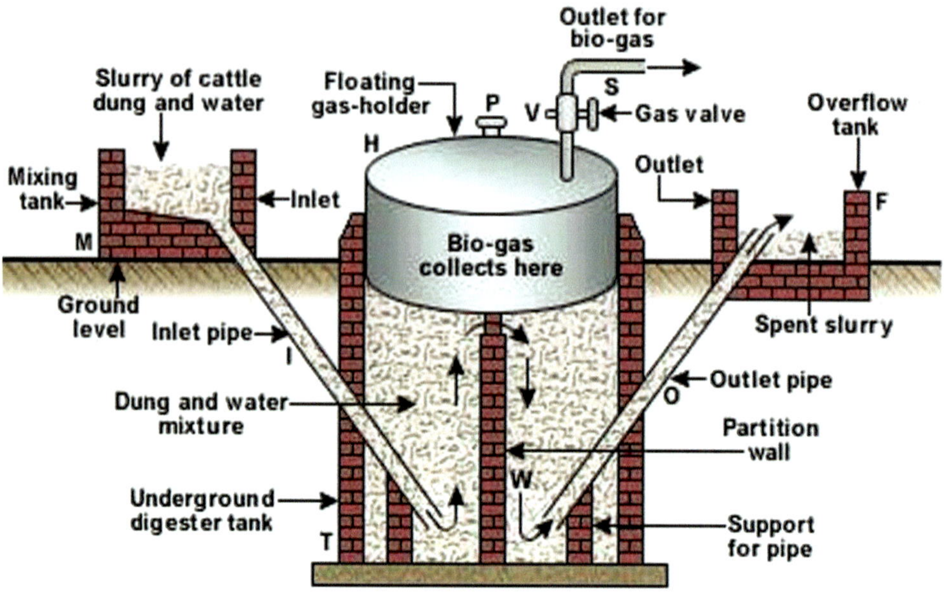

Floating gas-holder type bio-gas plant.

FLOATING DRUM MODEL

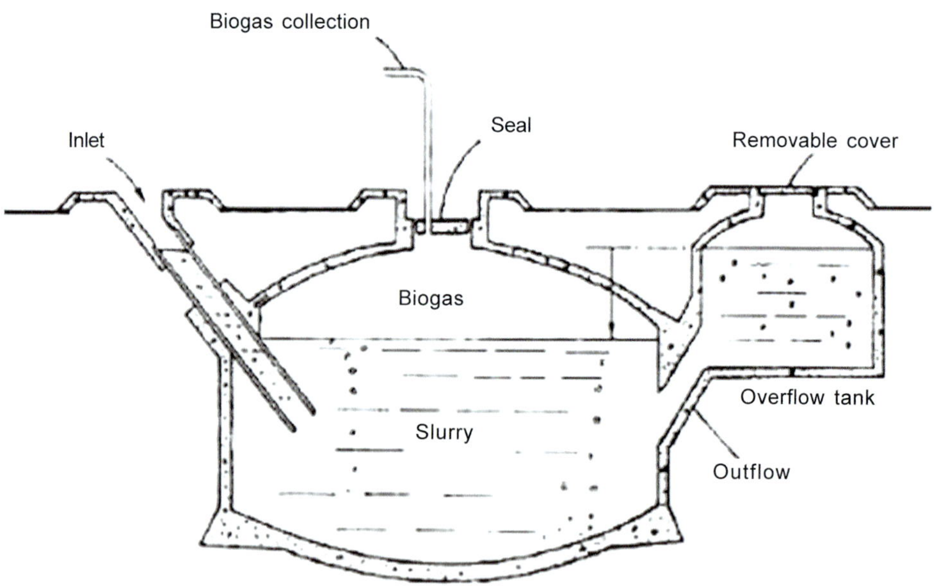

FIXED DOME MODEL

FIG. 4.14

TABLE 4.5 BIO-GAS PRODUCTION FROM DIFFERENT ANIMAL FEEDSTOCKS

Category of animal (feed stock)	Gas yield per kg of manure (m^3)	Average gas yield per animal/person per day (m^3)
Cattle	0.04	0.60
Buffalo	0.04	0.80
Pig	0.07	0.14
Poultry	0.06	0.009
Horse	0.04	0.60
Human	0.07	0.028

Jyothi *et al.* (2018) has reported that pig excreta has highest bio-gas production potential (0.092 ± 0.021 m^3) followed by goat pellets (0.088 ± 0.019 m^3), poultry droppings (0.082 ± 0.020 m^3) and cattle dung (0.077 ± 0.016 m^3).

Manurial value of bio-gas slurry

The biogas slurry is better than Farm Yard Manure (FYM) since it is well digested and has high nutrient content. It is considered more efficient and useful than compost as organic manure. The composition of bio-gas slurry and FYM has been worked out and furnished in Table. 4.6

TABLE 4.6 COMPOSITIONS OF BIO-GAS SLURRY AND FARM YARD MANURE (% ON DRY WEIGHT BASIS)

Constituents	Bio-gas slurry (%)	FYM (%)
Nitrogen (N)	1.41	0.78
Phosphorus (P2O5)	09.92	0.72
Potash (K2O)	0.84	0.65
Organic Carbon	27.32	24.4
C/N ratio	19.37	31.28

Compressed Natural Gas (CNG) Production

CNG is natural gas under pressure (100% methane), which is colourless, odourless and non corrosive. It can be used as an alternative to traditional petrol fuel for vehicles. Bio-gas is 60% methane. CNG can be produced from bio-gas. In this process, the contaminants of biogas are scrubbed out to produce pure methane and then high pressure is applied to compress it to form CNG which can be used as a fuel for automobiles and for cooking.

Disinfection and sanitation of pig houses

The disinfection of pig houses regularly is very important from the sanitary point of view to maintain the health and hygiene of pigs and to prevent the incidence and spread of various diseases. The manure should be removed twice daily. Scrape the floor and walls and clean it with pressurised water from pressure washers. The feeding trough should be swept thoroughly and scrubbed using hot water lye solution. Burn all scrapping and sweeping. If drinking water troughs are used, scrub it with disinfectant solution. Whitewash all walls, partition etc once in two weeks using 0.5 kg lime in 5 litres of water. Expose floors to sunlight.

CHAPTER 5

SWINE MANAGEMENT

Routine Piggery Operations

There are several operations that have to be carried out routinely in piggery farm just like on other livestock farms. The effectiveness of such operations will be more if they are carried out in an orderly manner as per a well prepared schedule.

TABLE 5.1 SCHEDULE OF DAY-TODAY OPERTIONS ON PIGGERIES

ApproximateTime (hours)	Farm operations
08.00-10.00	**A. ROUTINE DAILY JOBS**
	1. Cleaning of all pig sties.
	2. Cleaning of farm premises.
	3. Checking the herd for spotting out sick animals and their isolation
	4. When limited feeding is done, the breeding herd should receive its morning half of daily concentrate ration.
	5. Growing market hogs should be given free-choice concentrate ration in automatic feeders.
	6. Breeding gilts and sows should be put on the pasture before cleaning their pen.
	7. Isolation of "in-heat" sows and gilts and their breeding.

[Table Contd.

Contd. Table]

Approximate Time (hours)	Farm operations
	B. PERIODIC JOBS
	1. Periodical weighing of the stock.
	2. Cutting needle teeth of baby pigs.
	3. Castration and vaccination of stock.
	4. Identification of baby pigs by tattooing, ear-notching or tagging.
	5. Sale of market stock.
10.00-12.00	**A. ROUTINE DAILY JOBS**
	1. Disinfection of sties.
	2. Disposal/conservation of manure
	3. Sprinkling of pigs during summer
	B. PERIODIC JOBS
	1. Deworming
	2. Treatment of sick pigs for wounds, scabies etc.
	3. Preparation, storing and distribution of concentrate mixture
12.00-13.00	***Lunch-cum-rest period.***
13.00-17.00	**A. ROUTINE DAILY JOBS**
	1. Feeding the second half of daily concentrate ration to breeding and growing stock.
	2. Making daily entries in the farm records.
	3. Cleaning all sties.
	4. Transfer of near-parturition sows and gilts into farrowing pens.
	5. Checking nursing sows and gilts that are in the farrowing pen
	B. PERIODIC JOBS
	1. Weaning of piglings.
	2. Weekly scrubbing and white washing of the interiors of feed and water troughs
	3. Attending to farm purchases.

Pig Exercising

Sows which have sufficient exercise during gestation period usually have less difficulty at farrowing and they often produce pigs which are more vigorous than those produced by sows which did not have ample exercise. Nursing sows generally eat and lie down. Hence it is very important to provide some means whereby they will get exercise. One way to accomplish this is to feed the sows at some distance away from the sleeping quarters. Access to pasture, of course, is desirable for all classes of pigs.

Handling and Restraining of Pigs

Pigs should be handled very carefully especially boars and sows after farrowing, as they may be dangerous. Pigs can bite and they have complete set of teeth. Pigs have a tendency to bite any object thrown over it therefore a running noose or a twitch may be slipped over the upper jaw behind the tusk. As it is thrown, the pig bites the noose and it can be twisted tightly and held by an attendant.

Restraining pigs. A pig catcher is an instrument for securing a large pig. It is made up of iron rod 90 cm long with a handle at one end and a ring of 10 cm at the other end. The ring will be slipped down around the upper jaw behind the tusk (Fig. 5.1).

Small pigs may be caught by hind leg, seized above the hock or by their ears. If the base of the ear is held, the hand may slip owing to the greasiness of pig skin. Rubbing the hand in dry saw dust will give a better grip. It is not desirable to catch a pig by the tail alone. Sometimes the tail will tear causing blemish.

Large pigs can be secured by Barton's tether (Fig. 5.1). It consists of stout strap around the girth of the pig and another around the neck. Short-straps join these between the fore legs and on either side of midline.

Moving pigs: Pigs can be moved in the desired direction by entering the flight zone of the animal at 45°- 60° from directly behind the animal. Handlers should work on the edge of this zone to avoid extreme reaction. The simplest method is by using a small wooden board with a handle bar near the top. This board can be used to apply pressure to the sides, rear or front of the pig to indicate the direction you wish to take (Fig. 5.1). Excessive force should not be used to move a pig in any direction. However, pigs will refuse to move to a place that is dark as would be the case when moving from day light or a very poorly lit building or a vehicle or to climb steep ramps at the back of vehicles or a passage that is very narrow or with sharp bends.

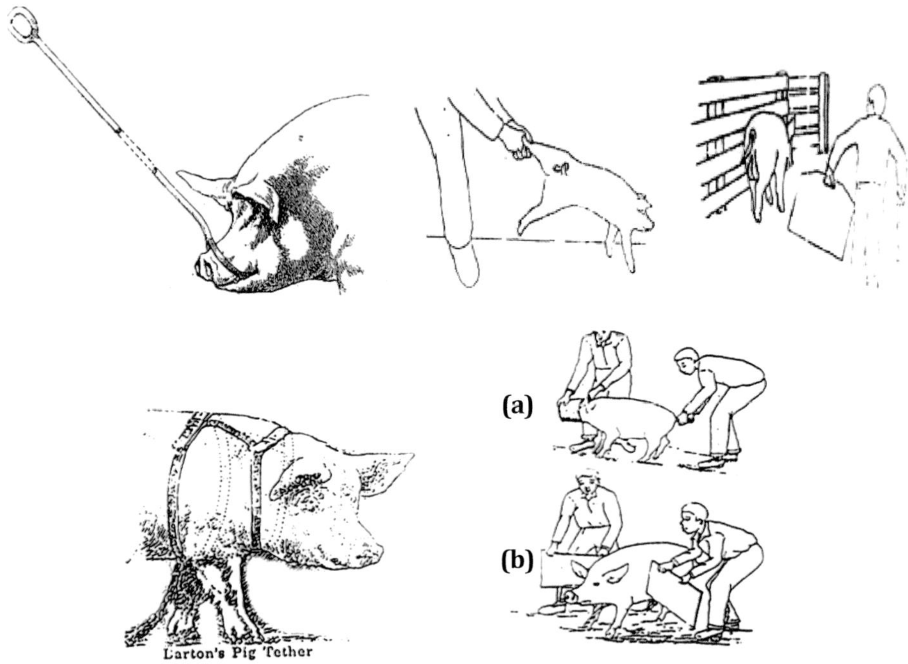

FIG 5.1 PIG CATCHER, MOVING PIGS, BARTONS TEATHER

Small pigs that weigh less than 15 kg can be caught by hand and moved to another pen or vehicle. If pigs have to be moved frequently between buildings, a permanent alley can be constructed to facilitate this process. Keep the alley narrow (30-40 cm), so that animal cannot turn around.

Backing of pigs: Sometimes it is necessary to back pigs out of a sty or crate. Use a solid object such as a bucket for backing the pigs (Fig. 5.1a). Position and hold solid objects immediately in front of the head or eyes. The pig will back up in its attempt to get away or keep moving a shovel or handle in front of the pig until the animal has moved from the crate (Fig. 5.2).

FIG 5.2 BACKING OF PIGS

Casting and securing large pigs: The animal's head is first secured by a twitch, running noose or with a pig catcher (Fig. 5.3 a, b). Side lines may be applied above the hock. Both the fore limbs are tied together and also both the hind legs. A long rope is fastened to the rope joining the forelimb and brought back between the hind limb. Another rope is tied joining the hind limbs and taken forward between the forelimbs (Fig. 5.3). Then the two longer ropes are pulled in front and behind. Then all the four legs are brought together and tied. 3 or 4 men may be required to cast a large boar or sow.

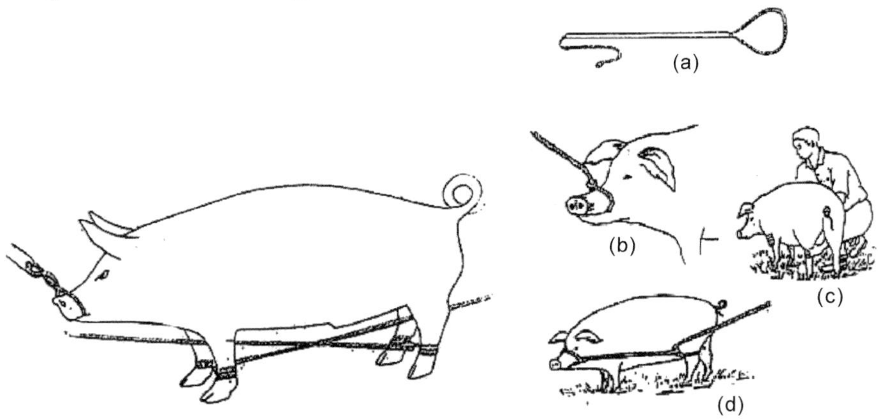

FIG 5.3 CASTING OF PIGS

Identification of Pigs

Identification is the process by which animals are marked for individual identity. Soon after birth all the animal should be marked properly by permanent identification.

Purpose of Identification

- It facilitates accurate recognition of animals.
- To differentiate between the animals of the same owner.
- Helps in avoiding substitution of one animal by another.
- For loan and insurance.
- Helps in keeping the production, breeding and pedigree record of the animal in a proper way.

Methods of Identification

Ear notching: It is the common method used for identification of swine. It is the process of removing a portion of the ear. This results in permanent identification. It is done soon after birth, so immediate identification is carried out and it results in less bleeding compared to older pigs. Older pigs are not ear notched because the process stimulates ear biting and cannibalism. This method comprises of cutting 'V' shaped notches at specified places along the border of ears by means of a short sharp scissors (Fig. 5.4). If the notch is too small, it is likely to close down and too large notch should also be avoided for fear of deforming the ear. The number should be recorded in a permanent record book preferably in two different books for reference. Before notching, one must decide on the number to be given to the animal. Small numbers being best. Also one must be clear on what notches must go on either or both ear, and what part of the ear they should go. The operator should keep clearly in mind which is the animal's right and left ear, regardless of whether he is standing before or behind the animal's head. The ears should be notched with the tips of the jaws of a pair of scissors or with a punch.

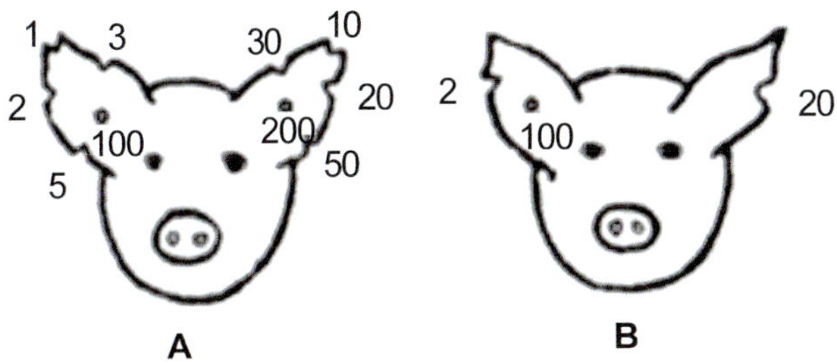

FIG. 5.4 EAR NOTCHING METHOD IN PIGS

Ear tagging: Ear tags can also be used as a method of identification of individual animals. Ear tags in breeding herds will greatly aid in keeping accurate swine records. It is the process of inserting a plastic, polyurethane or metal tag into pig's ear with a pliers type ear tag applicator. Ear tags should be inexpensive, permanent and easy to apply to the animal. The advantages are: 1.) It is a permanent method of identification 2.) By this method there is no ambiguity as there is only one pig with one number. While applying ear tags, properly restrain the animal, clean the ear with methylated spirit. Stand parallel to the animal and punch the back portion of the ear avoiding piercing the veins of the ear (Fig. 5.5).

FIG. 5.5 EAR TAGGING OF PIGS

Tattooing: It involves the placement of a marked tattoo into the skin of the animals. It is the most suited and efficient method of identification of newly born piglets. The process of tattooing involves punching of skin inside the ear with outline of desired number or letter with tattooing ink, normally black in colour. If tattooing is properly done, it will stay for a life time. The most important advantage of tattooing is that it is a permanent method of identification carried out at an early age. Disadvantages are 1. It cannot be done on dark coloured animals. 2. It cannot be read without close inspection of the animal. 3. Marks are not easily legible when compared or imprinted on pigmented skin.

Procedure for tattooing: The animal has to be restrained; the part to be tattooed should be thoroughly washed with soap and water and swabbed with methylated spirit to remove the grease. Tattooing ink is liberally applied over the part. Then the desired number or figure fixed to the tattooing forceps is first tested on a piece of hard paper and then firmly imprinted on the part by pressing the jaws of the forceps firmly. Some more ink is rubbed into the tattooed puncture with the bulb of the thumb. There may be slight bleeding in some areas. If the vein has been punctured by bad placing it can lead to serious bleeding. Unsatisfactory results are most commonly caused by dirty equipment, dirty and greasy skin, bad ink, inadequate application of ink or a misapplication of the tattoo.

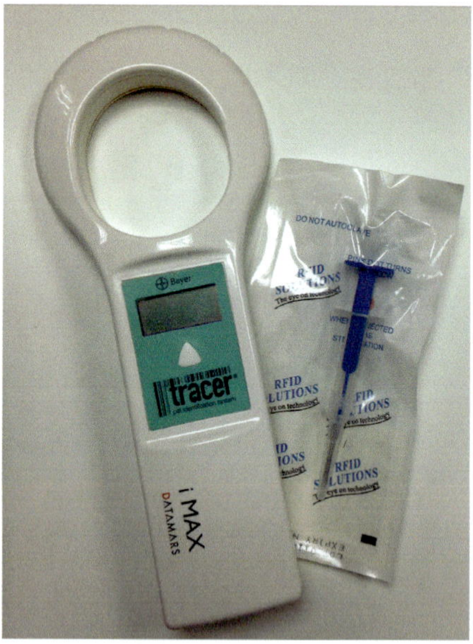

FIG. 5.6 MICROCHIP AND CHIP READER

Branding: Paint branding for temporary identification is the most common type of branding used in swine. Iron in the shape of numerals is used to paint or imprint a number on the back of the animal. Dyes like silver nitrate is use to brand new-born piglets, It is a temporary method of identification and the colour is retained only for two months. Hot iron branding is not used in swine as it will melt the subcutaneous fat.

Microchipping: This is the latest method of identification by inserting a micro-chip beneath the skin sub-cutaneously. It emits a radio frequency number which can be detected using a microchip reader. Microchip reader and chip are shown in Fig. 5.6.

Determination of age in pigs

Adult pigs have 44 teeth of these 12 are incisors, 4 are tusk, 16 are premolars and 12 are molars (with half of each type of teeth on each jaw). The tusk or canine teeth are more prominent in males than in females. The dental formulae of pigs are given below.

$$\text{Temporary dentition:}\left(I\frac{3}{3}C\frac{1}{1}PM\frac{3}{3}M\frac{0}{0}\right)2=28$$

$$\text{Permanent dentition:}\left(I\frac{3}{3}C\frac{1}{1}PM\frac{4}{4}M\frac{3}{3}\right)2=44$$

Temporary teeth are milky white, small in size and sharp. The tooth is made of enamel, dentine and pulp. Each tooth presents a part above the gums called the crown, coated with enamel. The part below the gums is the root which is embedded in the alveolus and is coated with cement. Between the two is the neck. Dentine forms the bulk of the tooth. It is hard and yellowish white in colour. Enamel covers the dentine of the crown and is bluish white. Cement covers the dentine of the root. The root is firmly attached to the alveolus by a vascular layer of connective tissue known as the alveolar periosteum. Pulp is a soft, gelatinous tissue in pulp cavity, well supplied with blood vessels and nerves.

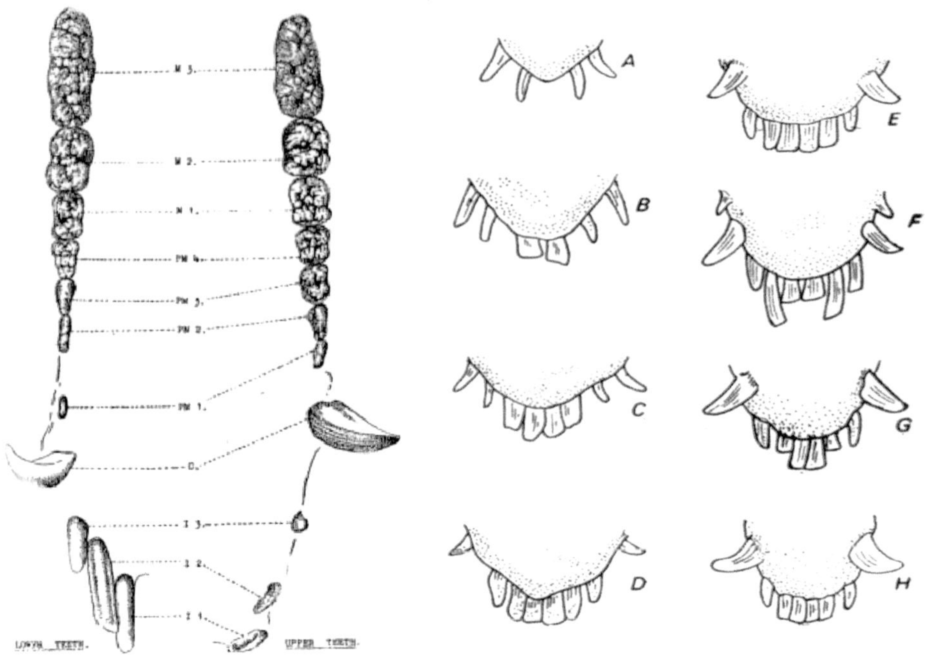

FIG. 5.7. DENTITION OF PIGS AND ERUPTION PATTERN OF TEETH

Incisors (cutting teeth) they are implanted in the alveoli of the pre-maxilla and mandible.

Canine teeth: (fangs or tusks) are situated little backwards.

Premolars and molars: Forms the sides of the dental arch. The first premolar is also referred to as 'Wolf teeth'.

The surface of the teeth that oppose the one on the other jaw is the table surface of the teeth. The surface facing the lips is the labial surface and that facing the tongue is the lingual surface.

Determination of age is a matter of noting the time of appearance and the degree of wear of temporary and permanent teeth.

Eruption of teeth: Eruption of all temporary teeth is completed by 5 months of age. Eruption of the pig's teeth is in many ways peculiar.

Incisors: The incisors do not erupt according to their position on the jaw. The corners erupt first by 6 months followed by the centrals and lastly the laterals. The same pattern is followed for wearing also. At the age of 6 months, the corner temporary incisors are replaced by permanent teeth (Fig 5.7 E). The permanent centrals (Fig 5.7 F) and permanent intermediaries (Fig 5.7 G) will replace temporary teeth by one and one and half years respectively. By this age all the permanent teeth will have erupted (Fig 5.7 H)

Canine: Canine tooth arises by 9 months.

Molar: The first tooth in each jaw immediately behind the canines erupts in 5 to 6 months and remains throughout the life. The next three premolars are represented in temporary dentition; the permanent ones replacing the temporaries in the normal way. The last three cheek teeth in each jaw are normal permanent molars. At 6 months, the corner temporary incisors are replaced by permanent teeth. The permanent centrals and permanent intermediates will replace temporary teeth by 1-$1^1/_2$ years. By this age all permanent teeth will be present. The eruption and replacement pattern of temporary & permanent teeth is given in Table. 5.2.

Smith field's rule

1. Pigs having their corners permanent incisor cut are considered more than 6 months of age.
2. Pigs having permanent tusks, more than half up are of more than 6 months of age.
3. Pigs having their central, permanent incisors up and any of the 1st 3 premolars cut are more than 12 months.
4. Pigs having their lateral, temporary incisors shed and the permanents appearing are more than 15 months.
5. Pigs with lateral, permanent incisors fully up, will be considered more than 18 months of age
6. Deciduous 3rd incisors and canines are erupted at birth, project laterally and have sharp points. They are called needle teeth. They are used for establishing 'Teat order or teat hierarchy' these tooth are often cut to prevent injury to the sow and other piglets (Fig. 5.9).

TABLE 5.2 ERUPTIONS AND WEAR OF TEETHS IN PIGS

Time of Eruption	Incisors			Canines	Cheek Teeth						
	1	2	3		1	2	3	4	5	6	7
Birth			O	O							
1 month	O		O	O		O	O	O			
2 to 3 months	O	O	O	O		O	O	O			
5 months	O	O	O	O	●	O	O	O	●		
7 to 8 months	O	O	●	O	●	O	O	O	●		
9 months	O	O	●	●	●	O	O	O	●		
1 year	●	O	●	●	●	O	O	O	●	●	
1 .2 years	●	O	●	●	●	●	●	●	●	●	
1.4 to 1.5 years	●	●	●	●	●	●	●	●	●	●	●

O-Temporary teeth ●-Permanent teeth

Castration of pigs

The process of rendering an animal infertile by mechanical or surgical means is called castration.

Objectives:

- To eliminate undesirable males.
- To prevent indiscriminate breeding.
- To prevent boar odour in cooked meat.
- To develop pork of superior quality (marbling)
- It can be done as a treatment of accidental injuries, tumours of testis etc.

Age of castration: 2-3 weeks of age.

Method: Open method. (Operation method). It is of two types 1. Open open method and 2. Open covered/closed method. In pigs open covered/closed method is practiced in which the skin of the scrotum is opened and the testes along with the tunica vaginalis is ligated and removed. In pigs as the inguinal canal is wide, opening of the tunica vaginalis may lead to hernia.

(Closed method by using Burdizzo's castrator is not done in pigs since testes are placed posterior and are not pendulous.)

Procedure.

1. Identify and segregate the animals.
2. Prepare for the operation by sterilising the shed and the instruments.
3. Secure the piglets by its hind limb and hang it conveniently upside down or hold it with both fore and hind legs together or use a squeege (Fig. 5.8).
4. Prepare the site by washing the area with soap and water, remove the hairs by shaving.
5. Apply Tincture Iodine.
6. Local infiltration anaesthesia is performed using 2% Lignocaine hydrochloride, subscrotally and intratesticularly.
7. Make an incision on the scrotum, deep enough, extend the cut down to permit proper drainage.
8. Pull out the testicle along with its membrane. Apply two ligatures at a distance of 2 cm and transfix them.
9. Cut between the ligatures to remove the testicle. Observe for any bleeding. Apply Tincture Benzoin and then leave the cord.
10. Approach the next testicle through the same incision. The incisions are not sutured and left as such for complete drainage.
11. Apply Tincture Benzoin pads
12. Inject a long acting antibiotic and Tetanus toxoid.

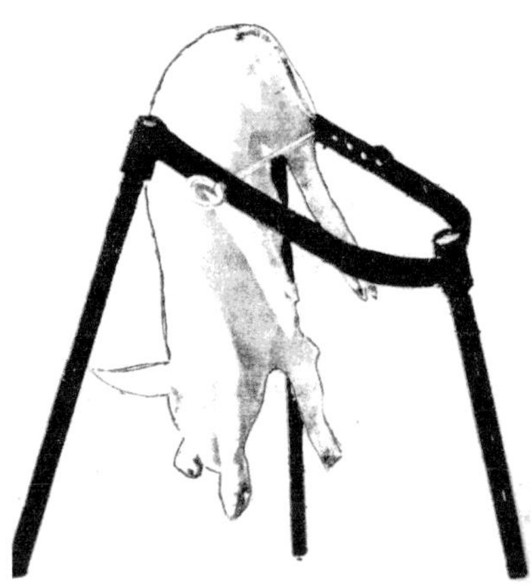

FIG. 5.8 SQUEEGE FOR PIG CASTRATION

Vasectomy: The procedure involves the ligation of Vas deference at two places and cutting and removing the central portion. Blood and nerve supply of testes are kept intact. Vasectomised boars are used to identity sows in heat. Such boars are called 'Teasers'.

TABLE 5.3 DIFFERENCES BETWEEN CASTRATION AND VASECTOMY

No	Particulars	Castration	Vasectomy
1.	Disposition of animals	More docile	less
2.	Silent heat in females	Not detected	detected
3.	Sex-libido	No	Possess
4.	Ejaculation	Does not occurs	Occurs
5.	Ejaculate	Does not occurs	Occurs without sperms
6.	Housed with other animals	Yes	No
7.	Tendency to fatten	Yes	No
8.	Testes	Absent	Present
9.	Size of scrotum	Shrunken	Full

Cross fostering

If a sow dies after farrowing or there is lactation failure due to udder trouble or the litter is larger than what a sow is able to raise, the result is as orphan pig. Two possibilities exist for the successful raising of such pigs: Adoption by a foster sow, or the use of either cow milk or a sow milk replacer. It is important that the newborn pig receives colostrums as the first feed. In pigs, the phenomenon of "gut closure" in which the absorption of large molecules such as gamma globulins of colostrums ceases following ingestion of feed is dominant. Therefore colostrums should be the first food to ensure immune globulin transfer to the blood. Such pigs from one litter may be usually transferred successfully to a sow with surplus teat, if done within 3 days of farrowing. The transfer should be made as early after farrowing as possible. It is advisable to soil the orphan pig with placental fluid from the foster sow so that sow accepts the orphan pig thinking as its own offspring. Transfer should be done as quickly as possible. Watch the sow to see if she accepts the piglet or not. In either case, an orphan piglet should get colostrums as soon as possible. Cross fostering is an important management practice for equalisation of the litter size.

Cow's milk is the best substitute for sow's milk. Butter milk or sweet skim milk can also be used. Each pigling may consume 300-500 ml milk per day. Best results may be obtained by feeding 5-6 times a day for the first few weeks. Thereafter, the frequency may gradually be reduced to 2-3 times. Any standard vitamin preparation 2 or 3 times the quantity used for infants may be administered to the piglings until they start taking feed. Milk replacer consists of one egg yolk thoroughly mixed with one litre of cow's milk. This mixture supplies a well balanced diet except for iron. To compensate for the lack of iron one eighth teaspoonful of ferrous sulphate may be added to one litre of milk.

Weaning of piglets

Reducing lactation length provides opportunity for increasing the number of litters per sow per year. If adequate litter size can be maintained, this creates the potential to increase the number of pigs weaned per sow per year.

Globally, the age at weaning varies between 2 and 8 weeks. The factors which influence the optimum age at weaning are 1.) The cost of feed ingredients for weaning diet, 2.) Cost of housing for piglets weaned at different ages. In developed countries, the age at weaning has been stabilised between 2 and 4 weeks. Whereas in countries where ingredients for early weaning diets are very expensive, age at weaning is likely to remain around 6 to 8 weeks. Under practical Indian farming conditions, piglets are weaned at 8 weeks (56 days) of age, when they are sufficiently grown to survive on feed alone. Piglets can also be weaned at 3 or 6 weeks of life called 'Early Weaning'. Recent trends are for early weaning of piglets; as approximately 4 weeks after farrowing the milk yield of sow becomes insufficient to support maximal weight gain of piglets. The advantages of early weaning are:

1. Sows are rebred earlier and swine producers can have more litters per year.

2. The body weight loss of sow during lactation is less, if early weaning is practiced, the sow can be sold at a better price.
3. It encourages the spreading of farrowing season. The requirement of piglets can thus be met all the year round.
4. A better control of disease in the herd is possible as it will prevent spread of disease from the sow to more susceptible young one.
5. Since there is a complete control of nutrients supply to the pigs in early weaning, piglets are more uniform, heavier and healthier at 8 weeks of age.
6. Pig losses due to crushing and savaging by the sow are avoided.

Disadvantages

1. There is an increase in the cost of feeding, since well balanced highly palatable fortified diets are needed.
2. The practice of early weaning requirc excellent know how and managemental practices.
3. Excellent sanitation is needed, otherwise losses due to disease may be high.

Split-Weaning

Permanent removal of part of the litter during a certain period before complete weaning is called split weaning or fractionated weaning. In the beginning, attempts to wean pigs early were not successful mainly because sucrose, which is not utilised by the new born pigs, was used as the energy source. As soon as it was realised that new born pigs can utilise lactose, it became possible to rear pigs without sow's milk. The teat order limits the growth of the smaller piglets as they are forced to feed on posterior teats which produce less milk. Thus the smaller piglets at birth remain smaller throughout the suckling period and later up to market due to the carry over effect. In order to overcome this and to have uniform crop of pigs, split-weaning is adopted. The litter is divided into two half *i.e.* the heavier half and lighter half. After 3 weeks of suckling by the sow, the heavier half is weaned on to a milk replacer and the lighter half will be with the mother. As the teat order is disturbed, the lighter half gets access to functional teats which produce more milk and compensatory growth occurs. The heavier half which is better adapted to the milk replacers also have good growth so that we get uniform piglets on weaning at 4 weeks. This method helps the piglets to overcome post-weaning growth lag and also helps to reduce the weaning-to-oestrous interval in sows. Lactational anoestrum is a prominent feature in swine and the sow generally comes to heat 4 to 10 days after weaning.

This is due to the fact that sucking by litter suppresses the ovarian activity through its effect on release of gonadotropic hormone from hypothalamus. This effect mainly depends on the sucking intensity (number of piglets with the sow) and the suckling interval (number of sucklings per day) both of which can be reduced by split-weaning. This helps the sow for an early and prompt post-weaning heat. This also reduces the body weight loss of sows during lactation. The sows may be re-bred on the first post weaning heat itself since this period is most fertile.

Pre-weaning mortality

Almost 61.33% of the deaths of piglets occur in the first week; hence good husbandry practices in early life will be profitable. The major causes of deaths were found to be agalactia, pneumonia and gastroenteritis. Strict hygienic measures may help in reducing the mortality due to infectious disease. Piglets with low birth weight were more prone to agalactia, trampling and infectious disease. Therefore, special attention to pregnant sow nutrition may reduce pre-weaning mortality.

Sow Productivity

The most appropriate criterion used to assess sow productivity is the number of piglets born and /or weaned per year per sow. This has two components; namely the number of piglets per litter and the number of litter per sow per year. The number of litter per sow per year can be increased by shortening the inter-farrowing interval, which inturn comprises of gestation length, lactation length and weaning-to-oestrous interval. As gestation length is a biological constant, reducing the lactation length and weaning-to-oestrous interval are the only means by which producers can increase the number of weaners per sow in a year besides reducing piglet mortality, which can be achieved by split-weaning.

Management of Different Categories of Swine

Care and management of pregnant pigs: The most important principle of caring for pregnant pigs is to provide them with a ration which will ensure the complete nourishment of the sow and her developing foetuses. In addition to proper nutrition, the pregnant animal should be provided with regular exercise. Forced exercise may be brought about by feeding some distance away from thc house or inducing the animal for moderate walk. Avoid overcrowding, mixing of new and old stock, slippery ground and over exciting of pregnant pigs. The important management practices to be followed are:

1. Calculate the expected farrowing date from the recorded breeding date to ensure that the sow is at the right place at the right time for farrowing.
2. Provide a separate clean and dry sty with non-slippery floor.
3. Provide well balanced pregnancy ration for uniform litter and better birth weight. Leguminous greens like berseem, lucerne and cow pea should be liberally fed. As major growth of the developing foetuses take place from 70-90 days of gestation, increase the amount of feed during this period.
4. Avoid overheating and provide plenty of fresh drinking water. Pregnant gilts/ sow should not be mixed with other animals to avoid fighting, which may lead to skidding and abortions.
5. Provide a bedding of 8-10 cm of chopped straw under the covered area.

6. De-worm the pregnant females 2 weeks before farrowing with broad spectrum anthelminthics like Albendazole, Fenbendazole at the rate of 10 mg/kg body weight. Do not excite the animals and keep them kindly treated.

Care of sow at farrowing time: Good care and management of the sow during farrowing can make the difference of the survival of 8-10 piglets out of the 12 farrowed. Approximately 30% of the pigs farrowed never reach weaning age and an additional loss of 5% occur after weaning. This means that only 65% of the pigs farrowed reach market age. This stresses the need for good care and management of sow before, during and after parturition.

1. Clean and disinfect the farrowing sheds and keep it vacant for a week. Cleaning should be done by scrapping, high pressure cleaners, steam cleaners and with a stiff brush.
2. Disinfect with commercially available disinfectants like quaternary ammonium compounds, iodoform compounds, lye or with phenolic compounds.
3. The pregnant females should be de-wormed 2-3 weeks before farrowing. She should also be treated for external parasites prior to admitting into the farrowing pen (0.01-0.05% of Deltamethin, Cypermethin or Amitraz).
4. Before moving into the farrowing pen, the sow should be thoroughly scrubbed with soap and warm water especially in the region of the sides, udder, under surface of the body and interdigital space. This removes dirt and adhering parasite eggs and bacteria that are potential diarrhoea causing agents.
5. Trim the overgrown long toe of sow to minimise crushing injuries.
6. Move the clean animal to the clean pen 10 days before farrowing. Provide light bedding of chopped straw 2-3 days before farrowing.
7. Reduce the ration by $^{1}/_{3}$rd till farrowing. Provide bulky ration. Withdraw feed 12 hours before farrowing. Provide access to fresh drinking water.

8. Keep the sow comfortable at a temperature below 24 °C. The farrowing pen should be dry, warm and clean and should offer protection from heat, cold and winds. Appearance of milk in teats when pressed indicates the approach of farrowing time.

Care of sow after farrowing

1. Clean the sow with lukewarm water and allow her to suckle her young ones. If there is no letting down of milk give oxytocin injection.
2. Provide sow her first meal 12 hours after farrowing. The sow may be fed the same bulky ration fed before farrowing. It is best to feed half of the ration on the first day and to increase the ration by 200 to 300 g daily, until she is full fed.
3. Check the rectal temperature. If it is above 40.2 °C gives appropriate treatment (Normal temperature after farrowing is 39.5 °C). Mastitis, Metritis, Agalactia syndrome (MMA) is a complex condition in which there is inflammation of the udder, uterus and complete lactational failure. MMA syndrome can involve metabolic, bacterial and hormonal factors, with stress playing a part. Since the main effect is loss of milk in the first three days post farrowing, the condition contributes to piglet losses from starvation. The symptoms are anorexia, lethargy, disinterest towards piglets, high fever, swelling of udder which becomes hard to touch, agalactia and mucopurulent vaginal discharges. In many cases, elevated body temperature is associated with this condition. So regular monitoring of sow's rectal temperature in the first 2-3 days after farrowing is important for diagnosis. A course of treatment with antibiotics and oxytocin is found to be effective. Heat stress to the sow due to artificial brooding may be another reason. Therefore provide cooler condition for sow. The udder and teats of sow should be dry and kept hygienic to avoid such problem.
4. The litter size should be equalised. Move piglets from large litters to small litters by cross fostering in order to make the litter size equal. This should be done during the first three days of farrowing.

5. Swab the sow's udder with a saturated solution of ferrous sulphate, zinc sulphate and copper sulphate to prevent nutritional anaemia in piglets.
6 Provide adequate nutritious balanced ration to the lactating sow for sufficient milk production.

Care and management of lactating sows: The weight as well as vigour of the piglets depend on a large extent on the management of lactating sow. During lactation, the sow affords heavy nutrient loss through milk. Nearly 180 to 185 kg milk is produced during a lactation period of 56 days. The peak yield is attained at 3 weeks. Hence up to one month of lactation, high nutrient diet is essential. Potable drinking water should be provided *ad-libitum*. 'Split suckling' or 'split milk feeding' can be adopted to level up piglets. It consists of splitting up the litter into two groups based on their size and allowing the lighter weight group first to feed followed by the heavier ones.

Care and management of gilts and sow: Maturity of gilt is more important than its age.

1. Skip the 1st heat and breed the gilts in their 2nd or 3rd heat to attain large litter size. Flushing should be carried out 2 to 3 weeks before breeding.
2. Flush the gilts (feed liberally) to ensure a gain of about 500 g per day during 2 to 3 weeks before breeding. It helps to increase shedding of more number of ova and thus increase litter size. Heat period of sow lasts for 2 to 3 days. They should be bred on the 2nd day of heat (24 hours after the onset of heat symptoms. Preferably 2 services at 8-12 hours interval). Females not conceived will repeat heat at 21 days interval.
3. The date of mating should be recorded from which the expected date of farrowing is calculate keeping in view the gestation period of 114 days.
4. The breeding stock ration must have 16% crude protein, containing cereals and oil cakes, well balanced with minerals and vitamins.

Leguminous fodders like berseem, lucerne, cowpea *etc* should be freely made available.

5. Since high temperature is detrimental to the survival of ova, gilts and sows for breeding should be maintained at a lower temperature of 16 °C.
6. The house should be clean with a controlled environmental condition to avoid stress.
7. Gilts and sow should be dewormed 2 weeks before farrowing.

Care of piglets at birth: The farrowing should be attended by an experienced person irrespective of the time of the day or night.

1. Remove the piglets soon after they are farrowed. Ensure that the breathing passage of all baby pigs is clear. In case of difficulty, swing the piglets at arm's length or give artificial respiration using a pig resuscitator. Clean all piglets free of mucous and make their body dry.
2. The navel cord should be cut at a distance 2.5 cm from the body with a sterilised scissor, apply a cord clamp and disinfect with tincture iodine.
3. If farrowing is abnormally prolonged, oxytocin may be used to speed the rate of delivery. Oxytocin should not be used if there is any suspetion that a piglet is blocking the birth canal. Continuous strong abdominal contractions for an extended period of time without the birth of piglets indicates the need for manual assistance. A well lubricated gloved hand should be inserted into the vagina; the piglet should be grasped gently and firmly pulled out. It is advisable to give antibiotic injection after this procedure to avoid complications.
4. Transfer the piglets to a creep area in which the temperature is 25 to 30 °C to protect from chilling.
5. Assist the piglets to get the teat and suck colostrum from their mother's udder. The sow should suckle the piglets 8 to 10 times initially in the first 24 hours period. Following parturition, sow will often call her litter to suck by emitting repeated short grunts and may emit loud barking grunts if an intruder disturbs the nest. The

sow rarely licks or grooms her young but sometimes appears to try to position the piglets near her udder or draw them towards her teats using her forelegs in scooping action.

6. Check the udder of the sow for agalactia, mastitis and number of functional teats to assess the nourishing capacity. Protect baby pigs from aggressive sow and from trampling.
7. As soon as the afterbirth is expelled, it should be removed from the pen and burned or buried in lime. This prevents the sow from eating the after birth. Remove wet, stained or soiled bedding and provide clean fresh material.
8. Provide micro-environment of light and heat to keep piglets comfortable. The temperature regulatory mechanism is ill-developed in new born piglets. When the new born piglets are exposed to cold condition, it has to mobilise limited glycogen reserve to maintain body temperature. Therefore, the baby piglet's body temperature falls very rapidly when environmental temperature falls. Normally they have sub normal temperatures during the first 30 minutes and it returns to normal during next 48 hours. The lower critical temperature is 34 °C for piglets. Therefore, for the new born piglets; temperature should be 32-35 °C under the infrared lamp. Infrared bulbs should be placed 45 cm above the pigs. After four or five days the temperature is lowered to 26-29 °C by raising the height of the lamp. The most effective method of reviving a chilled pig is to immerse the body except for the head in warm water. The pig should be kept like this for a few minutes, then removed and rubbed vigorously. Litter materials provide a temperature of 8 °C more to the piglets, which is a cost effective measure.
9. Provide supplementary feeding in case of inadequate nourishing ability of sow. In such cases, milk replacers can be fed using a soft plastic tube with a syringe attached.

Care and management of suckling pigs: The baby pigs are on their feet within minutes after birth and soon begin to explore the environment. Within a few minutes after birth, they soon locate the nipple and begin to nurse. On an average, they will have nursed successfully within 45 minutes following delivery. Because of this curiosity, the pen environment should be clean to reduce the risk of disease and parasites. The important management practices to be performed on piglets are:

1. The umbilical cord should be tied off to prevent blood loss and cut with a sterilised scissor at a distance of 2.5 cm away from the body. The cord should be dipped in a solution of tincture iodine. This will soon dry up and drop off, leaving a clean, non infected navel. A common malady, "navel ill," which causes lameness and death, may result from failure to disinfect the cord at birth.
2. The "needle" teeth should be clipped (Fig. 5.8). New born pigs have eight small, tusks like teeth called needle teeth or "wolf teeth". They are of no practical value to the pigs themselves and these sharp teeth often cause pain or injury to the sow particularly, if the udder is tender or cause injury to other pigs when they fight or play among themselves. Therefore, it is advisable to clip these teeth shortly after birth. A disinfected tooth cutting pliers may be used for this purpose. It is important to avoid loosening the root of the tooth, leaving sharp edges or causing injury to the gums.
3. The piglets should be identified immediately after birth by painting with silver nitrate solution and later by permanent identification methods like tattooing or ear notching. Ear notching is a simple and effective method of identifying individual pigs and litters and is

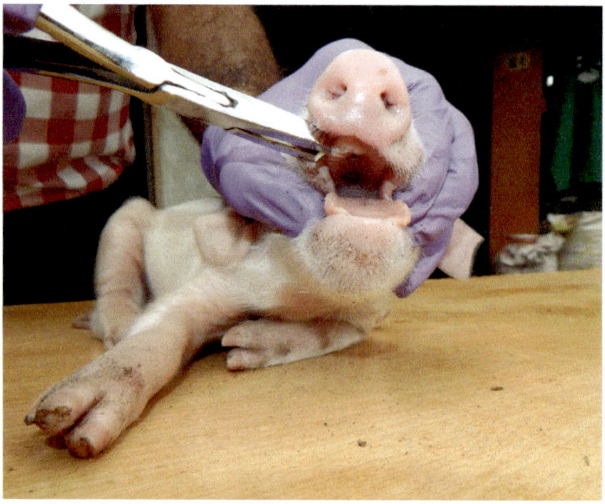

FIG 5.9 CLIPPING OF NEEDLE TEETH

essential for good record keeping. A disinfected commercially available ear notcher can be used or it can be done using a sharp scissor. Polyurethane ear tags are also used now-a-days for identifying pigs.

4. Management practices to prevent piglet anaemia should be adopted. Iron dextran injection should be given on the 4th and 14th day.
5. In totally confined commercial production system, docking of the tail is advisable to avoid the vice of tail biting. The tail should be cut as short as a stub as possible. Thermocautery scalpels can be effectively used for this procedure.
6. Piglets start nibbling feed as early as one and a half weeks. Creep feed should be supplied during the second week to encourage dry feed consumption. By this time baby pigs develop appetite for dry feed as their feed requirement begins to exceed the nutrition supplied by the sow's milk.
7. Watch the piglets closely for scours. Drugs given orally often work better than injections in preventing scours.
8. Excess male piglets not required for breeding are generally castrated at 2-3 weeks of age. This allows the piglets to recover from the growth lag before it receives weaning 'growths check.' At this age, the animal can be restrained easily and the stress is minimum.
9. The final task to be performed on suckling pig is weaning. Depending on the management status of the herd, conventional weaning at 8 weeks or early weaning at 4 weeks or split weaning before 4 weeks can be adopted.

Care and management of young stock

1. Record the birth weight and weekly body weight of the young stock to monitor growth and feed efficiency.
2. Culling and disposal of uneconomic animals can be carried out based on these records. Raise the piglets in small batches of 15-20 after weaning.

3. Deworm young piglets when they are five to six weeks old and again one month later using broad spectrum anthelminthics. Ectoparasitic infestation should be removed by spraying with synthetic pyrithroids or by giving intramuscular injection of dormectin.
4. Vaccination against important disease which causes huge economic loss to pigs has to be carried as per the vaccination schedule (Table 6.1).
5. Raise good virile piglets as boars and others should be castrated. The castrated animals are allowed to fatten well. Usually by 6 months of age pigs should attain a body weight of 60-70 kg.

Care and management of growing and finishing pigs: This period may be considered from weaning (9-10 kg) to the slaughter weight of 90-120 kg (Table 5.4). Entire males, castrates and females can be fattened for meat purposes. The entire males and females may have higher feed conversion efficiency than castrates. Castration if required may be done at the age or 2-3 weeks. Castrates are more docile and put on slightly more fat with marbling effect in pork.

1. Growers may be grouped according to sex, size and weight as uniformly as possible. The difference in weight between the small and large pig in a lot should not be more than 20%. Upto 15 pigs may be conveniently put together in a pen.
2. In summer, cooling devices may be provided in addition to shades.
3. Poor growers may be identified, culled and removed from the lot at the earliest.
4. Deworming is done after weaning and repeated once in 2 months.
5. If finished hogs are to be mixed from different pen for moving to market, this should be carried out in the morning when caretakers are present because the pigs will fight to establish a social hierarchy within the group.

TABLE 5.4 EXPECTED LIVE WEIGHT GAIN PER MONTH

Age in months	Live weight (kg)
1st month	4
2nd month	10
3rd month	20
5th month	50
7th month	85
9th month	120

Care and management of boars: Although boars reach sexual maturity at 7 months of age, it is generally recommended that they must be at least 10-12 months of age (100 kg body weight) before being put to regular use. Boars should be fed after service rather than before. Boar: Sow ratio may be 1:20. Outdoor exercise is good for keeping the boar thrifty and virile. He should neither be over fat nor run down. The boar may be handled firmly but gently, keeping in mind that he can be dangerous. Each boar should be housed separately and provided with 1.5 to 2.2 m^2 of sleeping area. Number of service for boar is 2-4 per week, depending on the age of the boar. For optimum performance, do not use young boar for more than one service daily and mature boar for not more than 2 services a day. Excessive service will result in the release of a decreased concentration of sperm as well as immature sperms. Bring sow to the boar at the time of service rather than allow the boar to run with the female stock all the time. A young boar turned with large number of sow may become shy. Young boars must be handled carefully. Hand mating is good for young boars. It is better to use young boars to experienced sow and experienced adult boars to gilts. Boars of the same age or size can be run together during the off-breeding season, but boars of different ages should not be kept together. It can result in fatal fighting. In hot weather, boars need to be kept cool with shades, fogging or sprinkling, or with a concrete hog wallow or evaporative cooling. Boars subjected to high temperature may demonstrate reduced fertility 4 to 6 weeks later. When temperature exceeds 27 °C mating should be carried out early in the morning or late in the evening. Avoid slippery floors in the breeding areas. Boars should be tested for brucellosis and leptospirosis periodically. Periodically carry out a fertility check by observing heat repeaters. If one out of 5 gilts returns to heat within 30 days, it reflects the low fertility of boars. Boars should be dewormed one week before breeding. Rotation of boars should be practiced to avoid in-breeding depression. If new boars are brought to the herd, he should not be used for mating until after a quarantine and isolation period of 30 days.

Ranting: Some boars pace back and forth along the fence often chomping their jaw and salivating. Such action is called ranting. Young boars that demonstrate excessive ranting may become "shieldy" (hard at the shoulders). This produces undesirable appearance although it may not affect their breeding. Isolating the animal from other boars and sow is usually effective in preventing this condition.

Boar taint: Meat from mature boars has an objectionable odour and flavour when cooked. This is associated with a fat soluble substance androstenone (5α-androst-16 ene-3-one) and skatol (3-methyl indole). Both androstenone and skatol are related in entire males and their concentration is very less in females. Boar odour is detected in most boars weighing more than 90 kg. Skatol is formed from tryptophan by gut bacteria and is taken up by fat tissues with the help of androstenone. Attempts to reduce this problem have been made by castration and also by implanting stilbesterol to boars.

Summary of good management practices

1. Feed and water should be protected from the contamination of manure and urine. Provide good drainage.
2. Carry out regular deworming of the herd. Undertake periodical spraying of insecticides to protect from lice and mange.
3. Different classes of pigs like young, old, weaners, breeders, pregnant stock and boars should be kept separately. Provide dry, clean, ventilated and draft free accommodation, which is very comfortable.

4. Prevent nutritional anaemia by providing iron to baby pigs.
5. Feed well balanced diet low in fibre. Provide green succulent fodders like berseem, lucerne and cow pea liberally.
6. Promptly segregate sick animal. Vaccinate against swine fever and hog cholera.
7. Follow a good calendar of daily operation.

Summer management of pigs

Among domestic animals, pigs have very poor heat regulating mechanisms. The heat tolerance capacity of pigs are very less due to the presence of relatively few sweat glands, reduction of melanin pigments in white coloured pigs, presence of sparse coat of hair and by the presence of a thick layer of fat below the skin around the body which acts as an insulator.

Therefore, in summer months exposure to direct sunlight for more than $2^1/_2$ to 3 hours will lead to a condition called Porcine Stress Syndrome (PSS) otherwise known as malignant hyperthermia or fatal syncope characterised by extreme muscling, nervousness, easily frightened tremors and skin blotching. PSS results in sudden death. Two different reliable tests are available to detect PSS. 1.) Creatine Phospho Kinase test (CPK test and 2). Halothane sensitivity test.

The creatine phosphokinase test (CPK test): Creatine phosphokinase is a serum enzyme that will be high in PSS swine. Therefore a high blood serum CPK level indicates that the animals are prone to PSS. This involves collection of blood on a card and sending it to the laboratory for the analysis of the activity of CPK.

Halothane test: This involves anesthetising pigs aged 3-11 weeks with halothane, (Fig. 5.9) to which PSS swine responds by showing extreme muscle rigidity within five minutes. There is a definite association between halothane sensitivity and the occurrence of Pale Soft Exudate (PSE) meat. Studies have shown that halothane sensitivity in pigs is caused by a single recessive gene, which suggests that selection using the halothane test would be an effective method of reducing the incidence of PSS and PSE. The halothane test is only capable of detecting the homozygous recessive animal which has received the halothane gene from both parents.

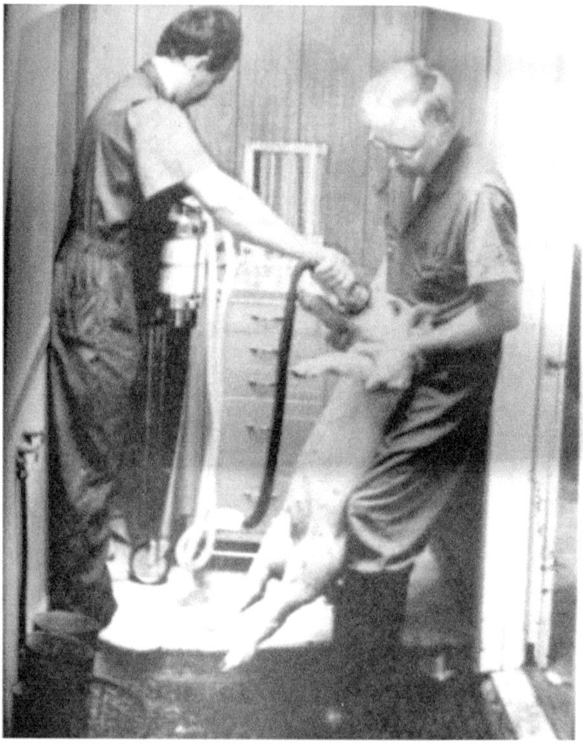

FIG. 5.10. HALOTHANE SENSITIVITY TEST

Methods to reduce heat load in pigs

Spraying devices: Spraying and sprinkling devices can be used in area where humidity is low to directly cool the animal. External sprinkling on the roof can be adopted to cool the whole house in high humidity areas. Internal sprinkling is the method of wetting down the pigs with a spray nozzle system under pressurised water (Fig. 4.1). When sprinklers are used one nozzle per 25 to 30 hogs are sufficient. Pigs can be wetted efficiently by spraying 2 litres of water per hog per day.

Internal fogging: In internal fogging, micro droplets of water which falls down from a pressurised nozzle cools the animal as it evaporates. Internal fogging does not wet the floors much as compared to spraying devices (Fig. 4.1).

Ventilating fans: To increase the rate of evaporative cooling, ceiling fans, exhaust fans and paddle fans can be used for forced ventilation. Movement of air produced by the fans will remove heat from the animals.

Air coolers: In low humidity areas, evaporative cooling devices like wet curtains and air coolers can be used.

Wallows: As mentioned earlier, hog wallows are the cheapest method to provide cooling for pigs. It is better to provide wallows only for breeding pigs as habitual wallowing in pigs leads to consumption of more feed.

Other Methods: Other methods to reduce the heat load of the shed are to paint the roof white on the exterior and black on the inside. Provision of insulating materials like thermocole at a distance of one foot below the roof, providing shades by planting trees around and planting grass around the sheds reduces reflected radiation.

Odour and fly control in piggery

Gasses and odours are given off from the animal manure by the breakdown of the organic part by anaerobic bacteria. Gases and odours can be reduced by mixing air with the manure. In liquid manure systems, this is done by agitators which forces air through the liquid. Odour control products may be used to control odours from animal manure. There are generally four types; masking agents, counteractants, deodorants and digestive deodorants.

Masking agents: Masking agents cover up the odour of wastes with the introduction of another odour. They are considered the most effective of the four types of odour control.

Counteractants: It attempts to neutralise the odour so that no odour remains.

Deodorants: Deodorants are chemicals that kill the bacteria that cause the odour. They are not as effective as the first two types

Digestive deodorants: These are bacteria that create a digestive process that eliminates the odour.

Handling and transportation of pigs

Pigs should be handled very gently; rough treatment may cause them to become very vicious. Permanent alleys around the sheds and ramps should be there in the farm for easy movement and for loading and unloading of pigs into vehicles. Simplest method of moving the pig forward is by using a wooden board with a handle near the top (Fig. 5.1). This board can be used to apply pressure on the sides, rear or front to indicate the direction of movement. However, pigs will refuse to move into dark place, climb steep ramps and negotiate sharp curves. Backing of pigs can be done by holding solid objects in front of the eyes like bucket or by moving a shovel in front of the pig (Fig. 5.2). Transportation of pigs can be carried out in trucks and in box trailers. The floor of the truck should be non slippery and appropriate bedding should be provided. It is better to transport pigs during the cool hours of the day. Ensure that the hogs are well watered before loading. High stocking density will result in death due to heat stress. Provide good ventilation and sprinkle water on the animals every 2 to 3 hours. Mixing of animals in the truck will result in fighting to establish social order. Start the journey after the dominance hierarchy has been established. Feed should be withdrawn 6 hours before transport.

Shrinkage of hogs

Pigs loose weight as they are transported to market. This weight loss is called shrinkage. The distance travelled is the most important criterion determining the amount of shrinkage. Shrinkage of about 2 per cent can be expected regardless how close the pig is being transported. This is caused by sorting, handling, loading and hauling. Pigs hauled 240 km or more may shrink as much as 4 per cent. Rough handling and a temperature above 20 °C will increase the amount of shrinkage. Careful handling of hogs reduces shrinkage losses.

Vices in pigs

Vices in pigs are of great economic importance as these seriously affect the growth and production of pigs.

Tail biting: Tail biting is very common in pigs due to the suckling instinct of pigs. Factors which precipitate tail biting are physical environment, stocking density, extremes of climatic variables, nutritional factors like deficiency of minerals and other diseases. The prevention is by docking the tail within the first few days of life.

Ear biting. Ears are bitten either at the base, where the ear joins the cheek or at the ear tip. The vice is usually noticed in intensive pig rearing. This vice is aggravated due to ear notching. Frequency of ear biting declines after 10-12 weeks of age.

Flank biting: This vice occurs in pigs kept in groups of 20 or more and from $1^1/_2$ to 5 months of age. The sites mostly involve lower rib cage and flank. This vice may be prevented by inserting rings in the dorsal area of the snout.

Snout biting: Snout biting is a vice in weaned pigs. The mutilation occures only when the pigs are kept in total darkness. Provision of light in the sty will avert this condition.

Pen fouling: Defecation of pigs in lying and feeding area is called pen fouling. Pigs usually defecate in the corner of the open area away from their feeding or lying area. The sub optimal physical environment with high stocking density may be the reason for this defect.

Naval sucking: Naval sucking can be a problem in early weaned pigs between 4 and 6 weeks of age. Persistent sucking of the naval may produce a fold of skin resembling a teat. The skin may become swollen, congested and painful. The control consists of separating the sucking piglet and the injured piglet. Early weaned pigs should be provided with attractive food with ample trough space.

Savaging: The savaging of the new born piglets by the sow which takes place during night may cause considerable financial loss to swine producers. The reason for savaging is not known but may be linked with pain at parturition and hysterical fear of the newborn piglets.

The piglet should be immediately separated and cross fostered or hand reared.

Aggressive Behaviour: Aggressive behaviour in groups of pigs is due to the fighting for the establishment of dominance hierarchy. It also depends on certain individuals.

Preparation of swine for shows

The steps in the preparation of swine for show are:

1. *Selection.* Choose the best type, which is true to the breed with good temperament, free from defect and with good growth as per its age and optimum condition of fleshing, two months before the show.
2. *Getling/taming.* Tame the pig well in advance using cane for guiding and training for walk and stop at will. Make it so gentle that it may be guided by just showing the cane or placing it along the side of the head.
3. *Trimming toes.* Cast the animal or else use pig catcher to keep the animal under control. Trim the toe by hoof trimmer and bring it into proper shape and size. Use hoof rasp to square up the soles, three or four weeks before the show.
4. *Removal of tusk.* Cast the pig and tie it to a post. Raise the upper jaw with pig catcher and remove the tusk by using a bolt clipper.
5. *Clipping hair.* Few days before the show, the hairs on the tail except switch, long hair over the hocks, body and belly of gilt must be clipped by using a clipper.

6. *Bathing*. Make the body wet with water. Apply soap and rub. Wash the body. Keep the animal in clean sty.
7. *Oiling*. Soak a small piece of cotton in oil and apply over the body and keep the pig in clean place. A further light application is desirable just before the show.
8. *Powdering*. Just before entering the ring, powder the body with corn starch to make white spots and dark hog very clear and white, pigs clean and attractive.
9. *Showing*. Keep the animal ready, well groomed and watered. Make the animal to walk before the judges. Keep the animal in good pose before the judges at a distance. Keep calm with no signs of nervousness. Prevent the animal from bite and fight with other hog.

RECORD KEEPING

The following records (5.5 to 5.9) are the minimum essential ones that are needed to be kept on pig farms.

TABLE 5.5 LITTER RECORDS

Breed________________________________Litter No.________________________________
(Notch, tattoo)

Data of Dam:

Pedigree________________________________SIRE________________________________
(Name, Reg. No., and ear notch)
DAM________________________________

Birth date________________________________
(Date and year)

Litter mate carcass data if any:

Carcass length ____________________ Av. Back fat____________Loin eye________
(cm) (sq.cm) (cm)

Sows________________________________Litter.
(1st, 2nd, etc)

Data on sire:

Pedigree________________________________SIRE________________________________
(Name, reg.No, and ear notch)
DAM________________________________

Birth date____

(Date and year)

Litter mate carcass data if any:

Carcass length ________________________ Av. Back fat____________ Loin eye________

(cm) (sq.cm) (cm)

Health services:

Date of swine fever vaccination___________ Date of erysipelas vaccination___________

No. pigs born:

Alive___________males;__________females. Date of deworming____________

Dead___________males;__________females.

Other, including iron injection ____

(list)

Mummies: Total________________males;_______________females______________

No. pigs weaned_________________males;_______________females______________

Individual Pig Record

Pig No	Sex	No. Teats Right	No. Teats Left	Birth Weight (kg)	Off colour Markings	Defects & Abnormalities	Weaning wt. at days
1	2	3	4	5	6	7	8

......Continued

Slaughtered Wt. At-days	Date Castrated	Date & Cause of Death	Disposal date and to whom?	Remarks
9	10	11	12	13

TABLE 5.6 BREEDING RECORD

Number of sows	Date of service	Number of Sire	Date to farrow
1	2	3	4

TABLE 5.7 FARROWING RECORDS

Number of sow	Date of farrow	No. of pigs born alive	No. of males in litter	No. of females in litter	No. of pigs raised	No. of pigs born dead
1	2	3	4	5	6	7

TABLE 5.8 FEEDING REGISTER (FOR THE MONTH OF ——————)

Date	Number of animals	Feed received (kg)	Issued for feeding (kg)	Balance in stock	Remarks
1	2	3	4	5	6

TABLE 5.9 HERD HEALTH REGISTER

Date	Animal No.	History	Symptoms	Diagnosis
1	2	3	4	5

...........Continued

Treatment	Result (Recovered dead / debilitated)	Name of attending Veterinarian	Cost of Treatment	Details of Vaccinations	Remarks
6	7	8	9	10	11

ECONOMICS OF SWINE PRODUCTION

Pigs are considered efficient converters of feed into meat. They grow fast. Ten sows and one boar can produce 160 to 200 young ones in a year that can be sold when they weigh around 65 kg. The dressing percentage in pigs is 65-70%, which means more meat is produced per animal than other species. The main cost is the feed. Swine being mono-gastric, needs high-energy concentrates, which are costly and not available in plenty. If waste foods like kitchen waste from hotels and hostels can be used for feeding, the feed cost can be reduced. Also an extensive system of rearing on cultivated tuber root crops can also reduce costs.

For bigger farms, one can develop statements of costs and returns based on the economic principles. However, pig keeping is considered an apt occupation for generating additional employment and income for the weaker section of the society, like members of backward and scheduled castes and scheduled tribes and other categories of rural poor. The Economics of 10 sow + one boar is worked out taking into consideration, the NABARD norms and the existing market rates and is described in Table.5.10

TABLE 5.10 ECONOMICS OF PIG FARMING (10 Sow + 1 Boar Unit)

Unit size	10 Sow with 1 Boar
State	Kerala
Unit Cost (Rs.)	7,48,892
Bank Loan (Rs.)	6,36,558
Margin Money (Rs.)	1,12,334
Repayment Period (Years)	5 with one year grace period
Interest Rate %	12
Benefit Cost Ratio at 15% Discount Factor	1.17:1
Net Present Worth at 15% Discount Factor	Rs. 3,40,701
Internal Rate of Return %	92

Technoeconomic Assumptions

Number of Sow (7-8 months old)	10
Number of Boar (8-9 months old)	1
Number of batches	2
Interval between two batches (months)	3
Number of farrowings per year	2
Average number of piglets per sow per farrowing	11
Mortality among piglets (weaners)	20%
Mortality among fatteners	10%
Weaning period (months)	2
SPACE REQUIREMENT (S.ft)	
Boar	70
Lactating sow with its piglets	100
Dry Sow	20
Fatteners (3-5 months age)	10
Fatteners (6-8 months age)	15
Store room	100
FEED REQUIREMENT (kg/day)	
Boar	3
Sow	3.5

Weaners	0.2
Fatteners (3-5 months age)	1.5
Fatteners (6-8 months age)	2
Concentrate feed % to total feed	20
Kitchen garbage % to total feed	80
CONSTRUCTION COST	
Cost of construction of sheds (Rs./S.ft)	200
Cost of construction of store room (Rs./S.ft)	350
PRICES	
Cost of breeding boar (100 kg) @ Rs. 90/kg	90000
Cost of breeding Sow (80 Kg) @ Rs. 90/kg	7200
Cost of Weaner feed (Rs./kg)	18
Cost of Grower feed (Rs./kg)	17
Cost of Adult feed (Rs./kg)	15
Cost of Kitchen garbage (Rs./kg)	1
Rate of Insurance (%)	6
Cost of medicines & vaccines	
Weaner/fattener (Rs./month)	8
Adult (Rs./month)	10
Other miscellaneous expense	
Weaner/fattener (Rs./month)	5
Adult (Rs./month)	3
Number of laboures required	1
Labour wages (Rs. Per month)	4500
No. of piglets sold per sow per farrowing (at 2 months of age)	8
No. of fatteners sold per sow per farrowing (at 8 months of age)	8
Sale price of piglets (Rs./Kg)	200
Average Weight of fatteners (Kg)	80
Sale price of fatteners (Rs./kg live weight)	70
Income from manure	
Weaner/fattener (Rs./month)	2
Adult (Rs./month)	5
Number of gunny bag per tonne of feed	13
Income from gunny bags (Rs/ bag)	8
Depreciation on sheds (%)	5
Depreciation on equipments (%)	10
Margin money (%)	15

Interest rate (%)	12
Repayment period (years)	5
Grace period (Years)	1

Fixed Investment/Construction cost

Particulars	**Specification**	**Physical units**	**Unit cost**	**Total cost**
Farrowing pen (4 Nos)	100 S.ft	400	200	80000
Boar cum service pen	70 S.ft	70	200	14000
Dry Sow pen (6 Nos)	20 S.ft	120	200	24000
Fattener Shed (3-5) months	10 S.ft	200	200	40000
Fattener Shed (6-8) months	15.5 S.ft	300	200	60000
Store room		100	350	35000
Installation of water supply system, well, electric motor, water tank	Lump sum			80000
Total construction cost				**3,33,000**
Equipment cost (Wheel barrow, shovel, heating lamp, pressure wash cleaner)	Lump sum			**40000**

Cost of breeding stock		**unit**	**unit cost**	**cost**
Cost of breeding Sow (7-8 m Av. Body Wt 80 kg	7200	10	7200	72000
Cost of breeding boar (8-9 m Av. Body Wt 100kg)	90000	1	9000	9000
Transportation charges @ Rs 100/ Animal		11	100	1100
Total cost of breeding stock				**82100**

TOTAL FIXED INVESTMENT **4,55,100**

Recurring Expenditure for one year	**Specification**	**Unit**		**Unit Price**	**Price**
Breeder feed cost					
Boar	3kg/Boar	3kg x 365 days	1095	1Rs./kg	1095
Sow	3.5 kg/Sow	3.5kg x 10 x 365	12775		
80% Kitchen Waste			10220	1	10220
20% Concentrate feed			2555	15	38325

Weaner feed cost 1st batch	0.2 kg/ piglet /day	0.2 x 110 x 56	1232	18	22176
2nd batch			1232	18	22176
Fattener feed cost (3-5 months)	1.5 x 80 x 90	10800			
80% Kitchen Waste			8640	1	8640
20% Concentrate feed			2160	17	36720
Fattener feed cost (6-8 months)	2 kg/ fattener/ day	2 x 80 x 90	14400		
80% Kitchen Waste			11520	1	11520
20% Concentrate feed			2880	17	48960
Total feed cost					**1,99,832**
Labour wages	Rs. 150/day	1		4500/ month	54000
Insurance cost					
Sow	@ 6% / annum	10	432	4320	
Boar	@ 6% / annum	1	540	540	
Insurance cost per year					4860
Cost of medicine & vaccines					
Breeder stock	11 No x 12 months	132	10	1320	
Weaner	160 No x 12	1920	8	15360	16680
Miscellaneous expense for breeder stock	11 No x 12 months	132	5	660	
Weaner	160 No x 12	1920	3	5760	6420
Electricity & water charges @ Rs.1000/ month		1000	12		12000
Total recurring expenditure					**2,93,792**

RECURRING EXPENDITURE	**1st year**	**2nd year**	**3rd year**	**4th year**	**5th year**
Feed cost	199832	199832	199832	199832	199832
Labour charges	54000	54000	54000	54000	54000
Insurance cost	4860	4860	4860	4860	4860
Medicines & Vaccine	16680	16680	16680	16680	16680
Miscellaneous expenditure	6420	6420	6420	6420	6420
Electricity and Water charges	12000	12000	12000	12000	12000
Total recurring expenditure	293792	293792	293792	293792	293792

TOTAL FINANCIAL OUT LAY (TFO) 7,48,892

BANK LOAN 85% OF TFO 6,36,558

MARGIN MONEY 15% OF TFO 1,12,334

INCOME

	No. of piglets	Av. Body Wt	Price	Income
By sales of piglets at 2 months of age	80	15kg	200/kg/ body Wt	240000
By sale of fatteners for slaughter	80	80kg	70/kg/ body Wt	448000
Manure adults	11	Rs. 5/month	660	
Weaners	80	Rs. 2/month	320	
Fattener	80	Rs. 3/month	1920	
Total				2900
Sale of gunny bags	134	Rs.8/bag		1072
				6,91,972

TOTAL INCOME

RECEIPTS	**1ST YEAR**	**2ND YEAR**	**3RD YEAR**	**4TH YEAR**	**5TH YEAR**
From sales of piglets	240000	240000	240000	240000	240000
From sale of fatteners for slaughter	448000	448000	448000	448000	448000
Sale of manure	2900	2900	2900	2900	2900
Sale of gunny bags	1072	1072	1072	1072	1072
TOTAL	691972	691972	691972	691972	691972
DEPRECIATION	1ST YEAR	2ND YEAR	3RD YEAR	4TH YEAR	5TH YEAR
Depreciation on equipment for 5 years (10%)	4000	4000	4000	4000	4000
Depreciation on buildings for 5 Years (5%)	49950	49950	49950	49950	49950
Total Depreciation	53950	53950	53950	53950	53950

Bank Loan and Repayment

Year	**Loan Amount**	**Interest**	**principal**	**Repayment**
1st Year	636558	76387	127312	203699
2nd Year	509247	61110	127312	188422
3rd Year	381935	45832	127312	173144
4th year	254623	30555	127312	157867
5th Year	127311	15277	127312	142589

Profit in 1st Year	
Income	691972
Bank loan	636558
Total Income	**1328530**
Fixed investment	455100
Recurring expenditure	293792
Depreciation	53950
Margin money	112334
Bank loan repayment	203699
Total Expenditure	**1118874**
Profit in 1st Year	**209656**
Profit per month	**17471**

Profit in 2nd year	
Income from sales	**691972**
Recurring expenditure	293792
Depreciation	53950
Bank loan repayment	188422
Total expenditure	**536164**
Profit in 2nd year	**155808**
Profit per month	**12984**

Profit in 3rd year	
Income from sales	**691972**
Recurring expenditure	**293792**
Depreciation	53950
Bank loan repayment	173144
Total Expenditure	**520886**
Profit in 3rd year	**171086**
profit per month	**14257**

Profit in 4th Year	
Income from sales	691972
Recurring expenditure	293792
Depreciation	53950
Bank loan repayment	157867
Total Expenditure	**505609**

Profit in 4th Year	**186363**
Profit per month	**15530**

Profit in 5th Year	
Income from sales	**691972**
Recurring expenditure	293792
Depreciation	53950
Bank loan repayment	142589
Total Expenditure	**490331**
Profit in 5th Year	**201641**
profit per month	**16803**

NET PROFIT

NON RECURRING EXPENDITURE	4,55,100
RECURRING EXPENDITURE	14,68,960
DEPRECIATION	269750
BANK LOAN	865720
MARGIN MONEY	112334
TOTAL RECEIPTS	40,96,418
TOTAL EXPENDITURE	31,71,864
NET PROFIT IN FIVE YEARS	**9,24,554**

CASH FLOW STATEMENT

Particulars	**1 Year**	**2 Year**	**3 Year**	**4 Year**	**5 Year**
Inflow					
From sale of animals	688000	688000	688000	688000	688000
From sale of manure	2900	2900	2900	2900	2900
From sale of gunny bag	1072	1072	1072	1072	1072
Total cash inflow	691972	691972	691972	691972	691972
Outflow					
On construction	333000				
On purchase of equipment	40000				
On purchase of animals	82100	0	0	0	0
On feeds	199832	199832	199832	199832	199832
On elctricity & water	12000	12000	12000	12000	12000

On labour	54000	54000	54000	54000	54000
On medicine	16680	16680	16680	16680	16680
On Miscellaneous expense	6420	6420	6420	6420	6420
On Insurance	4860	4860	4860	4860	4860
On depreciation	53950	53950	53950	53950	53950
On bank loan repayment		188422	173144	157867	142589
Total Out flow	802842	536164	520886	505609	490331
Net cash Inflow	**-110870**	**155808**	**171086**	**186363**	**201641**

CHAPTER 6

GENERAL SWINE HEALTH MANAGEMENT PROGRAMME

Many swine diseases are a result of complex interactions between infectious organisms, environment and management factors. Generally, economic well being of a swine operation depends on maintaining good herd health. Herd performance data such as growth rate and feed conversion are meaningful and sensitive indicators of health. Internal parasites, enzootic pneumonia and mild dysentery are examples of chronic disease that can reduce growth rates by 5-10% or more without causing death losses. Herd health security, the management of animal immunity and disease challenges within the herd, the monitoring and diagnosis of disease (both clinical and subclinical) and the treatment of sick individuals are the four main components of health management programme. A well planned herd health programme is aimed at prevention of disease and parasites rather than treatment. Management practices such as sanitation, isolation of new stock, selecting healthy breeding stock and proper care of sows and pigs are all part of good herd health management.

The general swine health management programme can be summarised as follows:

1. Maintain a closed herd *i.e.* as far as possible admit only specific pathogen free stock for new additions.
2. Don't allow other species to mingle with breeding stock.
3. Quarantine period should be provided for newly purchased animals at least 45 days before mixing with the herd.
4. Discourage or exclude non-essential persons and vehicles from the premises where pigs are bred and reared.
5. Follow the deworming and immunisation schedule regularly.
6. Keep fences, buildings and equipments clean, dry and disinfected.
7. Separate and maintain pigs by age group, keep adult breeders separate from weaned pigs and growers.
8. Provide adequate balanced feed for finishers and breeding stock.
9. Maintain the sty insect proof by spraying insecticides.
10. Prevent contamination of feed and water with urine and manure.
11. Clean the pens and houses before a new group of pigs are mixed.
12. Dead carcass should be properly disposed.
13. Inspect all pigs each day for signs of disease, watch for depression, lack of appetite, diarrhoea, lameness, swelling, prolapse, skin eruptions and signs of fever.

Deworming Schedule

1. Deworm the pigs immediately after weaning, followed by every month until six months of age, then once in every four months or after detection of ova in the faeces.
2. One month before breeding and expected date of farrowing.

Common Drugs Used

1. Levamisole hydrochloride – 8mg/kg bodyweight
2. Fenbendazole/Albendazole-10mg/kg bodyweight
3. Piperazine salts - 275-400mg/kg body weight

For mange infestation

1. Dormectin - 0.02mg/kg body weight Intramuscular (I/M) or 1ml/50kg body weight I/M.
2. Benzyl benzoate/deltamethrin/cypermethrin/amitraz for topical application.

Vaccination

Pigs have to be vaccinated against important swine disease which causes huge economic loss and for which there is no effective treatment. Vaccinate the animals as per the vaccination schedule.

TABLE 6.1 VACCINATION SCHEDULES OF SWINE

Disease and Vaccine	Age	Dose and Route	Duration of Immunity
Swine fever(Freeze dried vaccine)	3 to 6 weeks	1ml I/M	One year
Swine Erysipelas	2 months	1ml S/C	One year
FMD (Tissue culture vaccine)	2 months	2 ml S/C	One year
Brucellosis	2 months	2 ml S/C	Life long
Swine Influenza	At any stage	1 ml S/C	6 weeks

Source: Peters (2006)

Economically Important Diseases of Pigs

Anthrax

Causes: Bacterium: *Bacillus anthracis*

Symptom: Illness, fever, bloody diarrhoea, swollen neck, breathing difficulty, death.

Prevention: By vaccination with spore vaccine.

Arthritis

Causes: Bacterial or mycoplasmal infection; infected wounds or navels; poor hygiene.

Classification of Swine Diseases

Classification of swine dieases

Infectious

eg.

1. Swine Erysipelus-*Erysipelothrix rhusiopathae*
2. Tuberculosis-*Micobacterium tuberculosis*
3. Anthrax-*Bacillus anthracis*
4. Swine Plague/shipping fever-*Pasturella multocida*
5. Brucellosis-*Brucella suis*
6. Swine Influenza-*Influenza type -A*

Contagious

eg.

1.Hog cholera-*pestivirus*

2.Swine pox-*Vaccinia virus*

3.Foot and Mouth Disease-*Aptho virus*

4. Psudo Rabies-*Herpes virus*

5. Leptospirosis-*Leptospira pomona*

Parasitic

eg.

1. Ascariasis-*Ascaris suum*
2. Lung worm-*Metastrongyles* spp.
3. Kidney *worm-Stephanurus dentatus*
4. Nodular worm-*Strongylidae ransoni*
5. Mange-*Sarcoptes* spp. & *Demodex spp.*
6. Lice-*Haematopinus suis.*

Others

eg.

1. Piglet anaemia
2. Baby pig disease/hypoglycaemis
3. MMA ComplеX
4. SMEDI

Symptom: Hot painful swollen joints, lameness, reluctance to move.

Prevention: Disinfection and proper sanitation.

Actinobacillus Pleuropneumoniae (APP)

Causes: Bacterium: *Haemophilus pleuropneumonia*

Symptoms: Abdominal breathing, high fever, depression.

Prevention: Good sanitation and antibiotic treatment.

Atrophic rhinitis (AR)

Causes: Bacteria: *Pasteurella multocida, Bordetella bronchiseptica*

Symptoms: Sneezing, snuffling, nasal discharge, tear staining, twisting/ shortening of snout, poor growth.

Prevention: Good sanitation and antibiotic treatment.

Aujeszky's Disease (Pseudorabies, mad itch)

Causes: *Herpes Virus*

Symptoms: In pigs younger than 3 weeks, in-coordination, appetite loss, depression, vomiting, diarrhoea and convulsions may occur, then death of entire litters. Older pigs often have fever, appetite loss, sneezing, coughing, pneumonia, convulsions and sometimes blindness. Abortion, stillbirths and mummified pigs also can occur.

Prevention: modified live virus vaccine, recently released.

Baby Pig Disease (Hypoglycaemia)

Causes: Metabolic disease, low blood sugar due to inadequate food intake.

Symptoms: Shivering, weakness, failure to nurse, piglets weak and crying, death in 24 hours.

Prevention: Selection of breeding stock for higher milk yield, providing heat lamps, controlling litter size by redistribution of piglets.

Brucellosis

Causes: Bacteria: *Brucella suis*

Symptoms: Abortion as early as 17 days of gestation, posterior paralysis, lameness, infertility, still birth, premature birth and various reproductive problems.

Prevention: Isolation of infected animals. Eradication of total stock from the herd. Antibiotic treatment to prevent further complication.

Clostridial infection

Causes: Bacteria: *Clostridium perfringens* types A, B, or C

Symptoms: Diarrhoea, dysentery, death

Prevention: Disinfection and good sanitation.

Coccidiosis

Causes: One-celled parasites of the families *Eimeria, Isospora, Cryptosporidium*

Symptoms: Paste-like or watery diarrhoea, stunting, death in piglets.

Prevention: Good Hygiene and sanitation.

Colitis

Causes: Bacteria: *Spirochaetes, Yersinia, Clostridium, E. coli*

Symptoms: Diarrhoea, stunting.

Prevention: Good sanitation and antibiotic treatment.

Cystis/pyelonephritis

Causes: Bacteria, *e.g. E. coli, Eubacterium suis.*

Symptoms: Illness, pus in urine, weight loss, sudden death.

Prevention: Provision of potable water and good sanitation.

Diarrhoea-pre-weaning (milk scour)

Causes: Dietary problems, E. coli.

Symptoms: White, grey or brown scour, stunting, death.

Prevention: Good hygiene and sanitation.

Diarrhoea-post-weaning

Causes: Dietary problems; stress; mismanagement; virulent E. coli strains, Campylobacter or Salmonella.

Symptoms: Scours within 10 days of weaning.

Prevention: Good hygiene and sanitation.

Encephalomyocarditis (EMC)

Causes: *Encephalomyocarditis virus*, rodent infestation.

Symptoms: Sudden death, in-appetence, vomiting, staggering, breathing difficulty, stillbirths, mummies.

Prevention: Disinfection and good sanitation.

Enzootic pneumonia (EP)

Causes: *Mycoplasma hyponeumoniae.*

Symptoms: Cough, poor growth, death.

Prevention: Good hygiene and sanitation.

Eperythrozoonosis (EPE)

Causes: Bacterium: *Eperythrozoon suis, biting insects.*

Symptoms: Pale skin, fever, jaundice, poor growth, infertility

Prevention: Good hygiene and control of insects.

Erysipelas (Diamond skin disease)

Causes: Bacterium: *Erysipelothrix rhusiopathiae*

Symptoms: Fever, sudden death, red skin "diamonds", blue extremities, lameness.

Prevention: is by vaccination of hogs at six to eight weeks of age.

Exudative Epidermitis (Greasy pig Disease)

Causes: Bacterium: *Staphylococcus hyicus*

Symptoms: Brown greasy area of skin, scales, scabs, weight loss

Prevention: Disinfection.

Foot-and-mouth disease (FMD)

Causes: *Picorna Virus*

Symptoms: High temperature, lameness, blisters on tongue, snout and feet.

Prevention: Vaccination.

Glassers disease

Causes: Bacterium: *Haemophilus parasuis*

Symptoms: Fever, in-appetence, arthritis, breathing difficulty, cough, nervous signs, death.

Prevention: Good hygiene and sanitation

Inclusion body rhinitis

Causes: *Porcine cytomegalovirus.*

Symptoms: Sneezing, fever, infertility, small litters, still birth, mummies.

Prevention: Good hygiene and sanitation.

Japanese encephalitis

Causes: *Japanese encephalitis virus*

Symptoms: Sudden death, fever, nervous signs, abortion, mummies, still births, weak piglets, boar infertility.

Prevention: Good hygiene and control of mosquitoes.

Leptospirosis

Causes: Bacteria of the *Leptospira interrogans* group.

Symptoms: Abortion, infertility, mummies, still births, jaundice.

Prevention: Vaccination. Two initial vaccinations at breeding time establish protection throughout gestation.

Meningitis

Causes: *Streptococcus suis, Haemophilus parasuis, Salmonella etc.*

Symptoms: In-appetence, trembling, twitching, abnormal eye movements, convulsion, death.

Prevention: Good hygiene and sanitation.

MMA (Mastitis, Metritis, Agalactia)

Causes: Mismanagement, infections.

Symptoms: within 48 hours of farrowing: in-appetence, hungry piglets, fever, hard udder, agalactia, mucopurulent vaginal discharge.

Prevention: Prophylactic prepartum medication of long acting penicillin.

Mycotoxicosis

Causes: Fungal toxin in feed.

Symptoms: In-appetence, death, hyperoestrogenism, infertility, small litters.

Prevention: Sun dry feed ingredients. Protect feed from moisture.

Necrotic Enteritis (NE)

Causes: Bacterium: *Lawsonia intracellularis.*

Symptoms: Diarrhoea, wasting, pallor, black dung, sudden death.

Prevention: Good hygiene and sanitation.

Nipah Virus Disease

Causes: *Nipah virus*

Symptoms: Coughing, fever, respiratory distress, off feed, trembling, head pressing, spasms, leg weakness, abortion, still births, death.

Prevention: Good hygiene and sanitation

Parvovirus Disease

Cause: *Porcine parvovirus*

Symptoms: Delayed returns, not-in-pig, mummies, stillbirths

Prevention: Good hygiene and sanitation.

Porcine Dermatitis and Nephropathy Syndrome (PDNS)

Causes: Suspected Circoviral infection.

Symptoms: Multiple flat, red-blue, papules with necrotic lesions. Systemic necrotising vesiculitis and exudative proliferative glomerulonephritis.

Prevention: Good hygiene and sanitation.

Porcine Intestinal Adenopathy (PIA)

Causes: Bacterium: *Lawsonia intracellularis*

Symptoms: Diarrhoea, wasting, pot belly

Prevention: Good hygiene and sanitation

Porcine Stress Syndrome (PSS)

Causes: Genetic; stress & transport; heat stress

Symptoms: tremor, rigidity, sudden death, PSE meat, malignant hyperthermia, back muscle wasting.

Prevention: Isolation and culling by halothane sensitivity test.

Post-weaning Multisystemic Wasting Syndrome (PMWS)

Causes: *Porcine circovirus*

Symptoms: Wasting, poor growth, breathing difficulty, diarrhoea, jaundice, death.

Prevention: Good hygiene and sanitation.

Porcine Respiratory Corona Virus (PRCV)

Causes: Virus related to Transmissible Gastro Enteritis (TGE).

Symptoms: often no signs, sudden death.

Prevention: Good hygiene and sanitation.

Post-weaning Respiratory Disease Complex (PRDC)

Causes: Any combination of: PRRSV, mycoplasma, influenza virus, PRCV and management/environmental factors.

Symptoms: Coughing, difficult breathing, deaths.

Prevention: Good hygiene and sanitation.

Proliferative Haemorrhagic Enteropathy (PHE)

Causes: Bacterium: *Lawsonia intracellularis.*

Symptoms: Sudden death, black or blood-stained dung.

Prevention: Good hygiene and sanitation.

Porcine Respiratory and Reproductive Syndrome (PRRS)

Causes: strains of porcine reproductive and respiratory syndrome virus (PRRSV).

Symptoms: Illness, conjunctivitis, in-appetence, mummies, still birth, abortion, pre-weaning mortality, cough, infertility.

Prevention: Good hygiene and sanitation.

Rotavirus

Causes: Many strains of *Rotavirus.*

Symptoms: In-appetence, diarrhoea, vomiting, stunting.

Prevention: Good hygiene and sanitation.

Salmonellosis

Causes: Bacteria of the *Salmonella* family.

Effects: Death, fever, diarrhoea, nervous signs.

Prevention: Good hygiene and sanitation.

Streptococcus suis infection

Causes: Bacterium: Streptococcus suis.

Symptoms: Arthritis, nervous signs, stunting, fever, death.

Prevention: Good hygiene and sanitation.

Swine Dysentery (SD)

Causes: Bacterium *Serpulina hyodysenteriae.*

Symptoms: Diarrhoea, with mucus and sometimes blood, weight loss, reduced feed conversion, death.

Prevention: Good hygiene and sanitation.

Swine Influenza/Hog flu.

Causes: Swine, avian and human influenza viruses.

Symptoms: Fever, inappetence, sneezing, coughing, low mortality.

Prevention: Good hygiene and sanitation.

Swine Fever/ Hog cholera

Cause: Swine fever virus.

Symptoms: Pyrexia up to 40-41 °C, dullness, anorexia, arching, partial paralysis, staggering gait. Button ulcers in large intestine and Turkey egg appearance of kidney are the pathagnomonic PM lesions.

Prevention: Vaccination at 3rd week and 3rd month.

Swine Plague

Causes: Bacteria. *Pasturella suiseptica.*

Symptoms: Spasmodic coughing, rapid breathing, difficulty in breathing, inflammation of mucous membrane of eyes, throat, loss of appetite and constipation followed by diarrhoea.

Prevention: Good hygiene and sanitation.

Swine Pox

Causes: *Swine Pox Virus.*

Symptoms: Fever, lack of appetite, eruptions over the skin commonly on ears, under surface of body and thighs, ulcerations in digestive tract.

Prevention: Good hygiene and sanitation.

SMEDI- Still birth, Mummification, Embryonic death and Infertility.

Causes: Complex disease, bacterial and viral causes.

Mainly gilts are affected.

Prevention: Only by managemental practices.

Transmissible Gastro Enteritis (TGE)

Causes: Corona Virus.

Symptoms: Pyrexia, sudden vomiting and diarrhoea, death

Acute inflammation of the intestine.

Prevention: Good hygiene and sanitation.

Vesicular Exanthema (VE)

Causes: Exanthema virus

Symptoms: Elevation of temperature followed by development of small vesicles on the snout, teats, udder, and feet causing lameness.

Prevention: Good hygiene and sanitation.

CHAPTER 7

PORK PRODUCTION TECHNOLOGY

Pigs which are the most efficient converters of feed into pork are reared only for meat purpose. Therefore, the scientific method of slaughter is of paramount importance in enhancing the acceptability of pork among the people. The scientific slaughter procedure is detailed herewith.

Ante-mortem inspection: It is the examination of food animals in holding pens within hours prior to slaughter for deciding the fitness to slaughter while at rest. Feed is withdrawn for 12-24 hours in the holding pen but ad-libitum water is provided. They should be well rested (at least 12 hours) prior to slaughter. In hot summer months sprinkling of water and environmental modification should be attempted to reduce stress. Pigs are most prone for heat stress.

Stunning: It is the humane method of desensitising the animal to the pain during killing. It is done using captive bolt pistol. In pigs the pistol is placed about 2.5 cm above the level of eyes and fired into the cranial cavity. It produces immediate insensibility by destruction of the cortex and deeper parts of the brain, changes in intracranial pressure and the sudden jerk producing acceleration concussion. Alternatively, electrical stunning (70 V for 2-3 min) can also be performed. After stunning the pigs are hoisted usually using a chain pulley for efficient due to gravity bleeding.

Sticking: By using a sticking knife, a small incision is made in the skin just anterior to the part of the breast bone. The knife is inserted above the breast bone to a distance of 12-15 cm into the jugular furrow and is given a quick turn to a quarter circle and withdrawn with a down ward motion. This will severe the carotid artery and jugular vein providing a thorough bleeding in a time of 5 minutes.

Scalding: It is the process of dipping the carcass in hot water at a temperature of 60-80 °C for 6 minutes to remove the hairs. The temperature of water can be checked with a thermometer. Pig carcass is dipped into the scalding tank or hot water is poured for 4-5 minutes till the hair pulls off the skin.

Scrapping: Pig carcass is removed from the scalding tank on to a platform for removal of hair with a bell shaped hog scraper or the carcass is elevated on to a de-hairing machine to remove hair mechanically. Remaining hair on the legs or body may be scrapped out using a knife. After scrapping, the shackles are hooked around the pastern and the carcass is hoisted usually with a chain pulley onto over head rails.

Singeing: It is the removal of fine hairs with a stream of fire from a blow lamp. It also burns the left over hairs or bristles on the body. Blow lamp can be LPG or kerosene operated. LPG lamp produces a fine blue flame and doesn't leave any smell. The carcass is then thoroughly washed with hot water and then with cold water and cleaned thoroughly.

Steps in Dressing of pigs

1. The head is removed by cutting through the atlanto-occipital joint.
2. Loosen the trachea, it is freed and tied with a string.
3. Remove the fore feet one inch above the knee and hind feet at the lower end of the hock.
4. Open down the midline on the ventral side from rectum to the ribcage, penis is loosened.
5. Split the breast bone and os-coxae (aitch bone).
6. Loosen the intestine, leave kidney fat intact.

7. Lift out the intestine, liver and stomach.
8. Split the body cavity inside and then forward along the diaphram
9. Lift out all the abdominal parts with upward movement.
10. With downward movement, pull the lungs and heart including aorta.
11. Pull the organs, wash and weigh.
12. Split the carcass down to the middle of the back bone using a splitting saw (Fig. 7.1).
13. Weigh the carcass of each side.

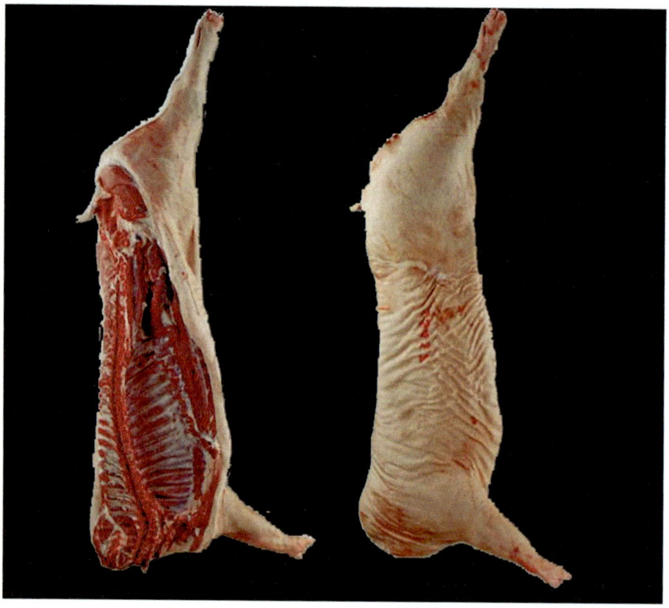

FIG. 7.1 SPLITING OF CARCASS

Cut-up Parts

The wholesale cuts of pork is shown in Fig. 7.2.

Rough Shoulder (Boston butt; picnic shoulder and jowl): Remove the rough shoulder by cutting in a line parallel to the head- lose limb and through the third rib from the front. A cut through shoulder joint is trimmed by cutting close to the neck line. Separate shoulder into picnic (lower 2/3rd) and boston butt (upper 1/3rd)

TABLE 7.1 SLAUGHTER DATA OF A TYPICAL LARGE WHITE YORKSHIRE

Slaughter data	
Species	Pig
Breed	Large White Yorkshire
Identification No	505
Sex	Male
Date of birth	01.07.2018
Age	1 year and 6 months
Live weight	150 kg
Dressed weight (hot carcass)	111.5 kg
Weight of blood	2 kg
Weight of head	10.2 kg
Weight of feet	2.8 kg
Weight of stomach and Intestine	11.98 kg
Weight of Liver	2.77 kg
Weight of heart	0.466 kg
Weight of trachea + lungs	0.818 kg
Weight of kidney	0.556 kg
Weight of spleen	0.30 kg
Weight of testicles	0.82 kg
Weight of hair	0.15 kg
Back fat thickness	First rib-3.3 cm
	Last rib- 1.0 cm
	Last Lumbar- 1.5 cm
Skin thickness	1.4 cm
Weight of tail	0.225 kg
Carcass length	92 cm
Dressing percentage	72%
Meat yield (%) based on hot carcass weight	79.74%
Meat yield (%) based on slaughter weight	57.41%
Bone yield (%) carcass weight	16.44%
Bone yield (%) based on slaughter weight	11.87%
Meat : Bone ratio	4.8 : 1
Fabrication loss	1.61

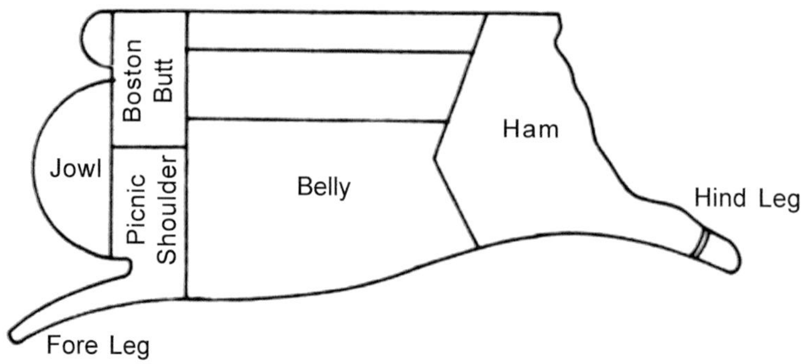

FIG. 7.2 WHOLESALE CUTS OF PORK

Ham (rough): Ham is separated from the loin by cutting in between 2nd and 3rd sacral vertebrae at right angle to the line of leg and regular ham is made from this rough ham by suitable trimming of the fat and lean on the exposed surface of the ham.

Loin: Loin is separated into upper loin, lower side and spares by a parallel cut to the back bone made at a point three fingers width between the vertebral column and the shoulder end and extending up to the leg.

Whole sale cuts and percentages

Jowl-3%	Side and spare ribs-17.3%
Boston butt-6.6%	
Picnic Shoulder-8.8%	{Bacon}
Loin-18%	Ham-21%

Back fat thickness and skin thickness are measured at the first and last rib and at the last lumbar vertebrae. Carcass length is measured from the first rib to the anterior part of aitch bone (Fig. 7.3).

Loin Eye Area (LEA): Loin eye area is measured between the tenth and eleventh ribs in swine (Fig. 7.4). Only the large longissimus dorsi muscle is included in the measurement. The muscle is traced on a clear piece of acetate paper and the area is later determined using a planimeter. It is the best indicator of the degree of muscling in the carcass. The loin eye area should be 4-5 square inch or more in meat-type hogs.

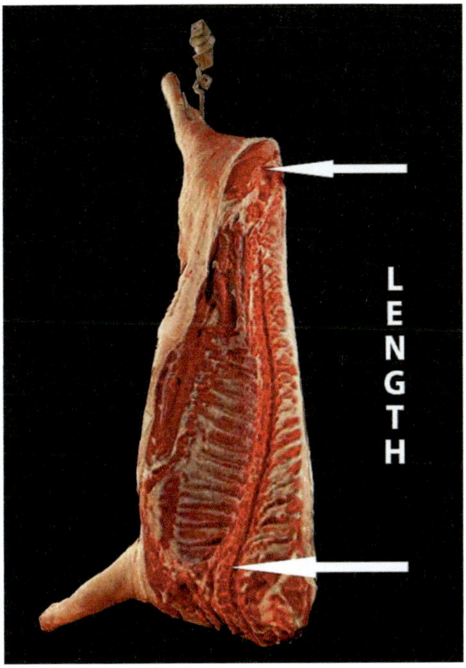

FIG 7.3 CARCASS LENGTH

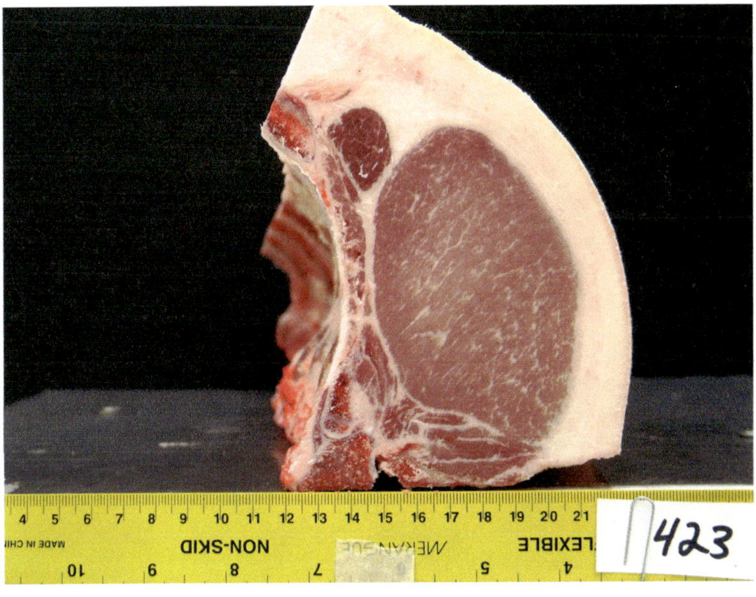

FIG 7.4 MEASUREMENT OF LOIN EYE AREA

TABLE7.2 WHOLE SALE CUTS DATA OF A 63.5 Kg LARGE WHITE YORKSIRE

Sl.No	Cuts	Left	Right	Meat		Bone	
				Left	Right	Left	Right
1.	Jowl	0.74	0.20	0.71	0.20	—	—
2.	Boston Butt	1.04	3.00	0.99	2.37	0.04	0.61
3.	Picnic Shoulder	3.55	3.30	2.11	2.59	0.69	0.70
4.	Loin	6.50	8.36	5.38	6.52	1.12	1.83
5.	Bacon	2.94	3.27	2.67	2.98	0.28	0.31
6.	Ham	6.01	5.82	4.94	5.00	1.08	0.86
	TOTAL	**20.75**	**23.95**	**16.8**	**19.66**	**3.21**	**4.31**

CHAPTER 8

BIO-FUEL PRODUCTION FROM PORK FAT

India with 16% of the world population could boast of only 0.5% of the world oil reserves. The estimated crude oil import cost comes to about 10% of the country's Gross Domestic Product. The current level of production barely caters to 26% of the petroleum demand. India imports 75% of its crude oil requirements. The crude oil price has been increasing and fluctuating causing geopolitical instabilities around the world. Besides the billions spent on importing crude oil, we are also importing billions of tonnes of CO_2 and other green house gases causing climatic changes as evidenced by changing rainfall patterns, rising sea levels and temperatures. The demand for diesel is five times higher than the demand for petrol in India and any increase in diesel price immediately pushes up inflation. Therefore, the government was forced to decontrol the diesel pricing as they no longer could bear the burden of fuel subsidies. With the Indian economy poised for a robust growth of 9 to 9.5% per annum, energy security has become the key issue in policy formulation and planning. Energy security would remain vulnerable until alternative fuels to supplement conventional fuels are developed based on indigenously available feedstocks. In bio-fuels, the country has a ray of hope. Biofuels are non-polluting and virtually in exhaustible. Biofuels can increasingly satisfy the energy needs in an environmentally benign and cost-effective manner while reducing dependence on the import of fossil fuels. The national bio-fuel policy-2018, proposed 10% reduction in the import of crude oil in 4 years time by blending 20% bio-diesel. Thus huge demand cannot be met from non-edible vegetable oil feedstock. Therefore, development and utilisation of new indigenous bio-mass feedstock and development of next generation of more efficient bio-fuel production technologies is the need of the hour. In this context, bio-diesel production from animal fats offers new scope as a potential means to stimulate rural development, lower emission of harmful pollutants and decrease green house gas emission, while contributing to national energy security by reducing dependence on oil imports and providing fuel with good fuel properties for the diesel engine.

Importance of Bio-diesel

Diesel engines are the major source of air pollutants released into the environment. The important pollutants from diesel engine are unburned carbon compounds, carbon monoxide and dioxide, nitrogen oxides, sulphur and polycyclic aromatic compounds. The diesel vehicles were banned in New Delhi for serious problem of air pollution due to higher emission of polluted gases. The acid rain, global warming, climatic changes and health hazards are the ill effects of increased polluted gases like $SO_{2,}$ CO and particulate matter in atmosphere.

Today's diesel engines require a clean burning stable fuel that performs well under a variety of operating conditions. Bio-diesel offers a number of environmental, social and economic advantages including

- Lower emission of harmful pollutants
- Decreased greenhouse gas emission
- Increased energy security
- Decreased dependence on oil imports and
- Good fuel properties for vehicle.

Bio-diesel has good fuel properties, comparable to or even better than petroleum diesel. It has 10 per cent built-in oxygen content that helps it to burn fully. It's cetane number (an indication of its fuel burning efficiency) is 52 for bio-diesel from Jatropha oil, 72 for bio-diesel derived from tallow and 72.5 for bio-diesel derived from chicken oil, higher than 54.4, the cetane number of most petroleum diesels. The esters of the long-chain fatty acids of bio-diesel are excellent lubricants for the fuel injection system. It has a higher flash point than diesel, making it a safer fuel. Other advantages are the almost zero sulphur content and the reduced amount of carbon monoxide, unburned hydrocarbons and particulate matter in the exhaust.

Animal Fats for Bio-diesel Production

Compared to bio-diesel produced from vegetable oil, animal fat based bio-diesel offers the advantage of reduced raw material cost, increased calorific value, cetane value (which improves engine performance) and greater oxidative stability which is due to the presence of natural antioxidants as compared to bio-diesel of vegetable origin. Animal fat bio-diesel also shows best performance as a lubricity enhancer for low-sulphur diesel.

Composition of Animal fats

The fatty acids in animals are mainly of medium to long chain length having 12 to 22 carbon atoms in their molecule, with the basic structure of CH_3-$(CH_2)_n$–COOH. The fatty acids from animal fats are about 40 per cent saturated; that is, two hydrogen atoms are attached on every carbon atom. About 40 per cent are monounsaturated fatty acid (MUFA) having one double bond and one hydrogen atom attached to adjacent carbon atom. A smaller proportion, (about 2% -25%) are poly unsaturated fatty acid, (PUFA) having more than one double bond.

One of the major fatty acids (contributing over 30% of total fatty acids) in all meat is Oleic acid (18:1 *cis*-9). The physical property of fatty acids such as melting point is influenced by the length, degree of unsaturation and configuration of the fatty acid molecule. The melting point of saturated and trans-fatty acids is higher compared to unsaturated and '*cis*' fatty acids due to long chain length and lower number of double bonds present in the molecule.

Rumen bio-hydrogenation of fats results in the production of saturated fatty acids and 'trans' fatty acids,which are more complex than those in non-ruminants.Varieties of fatty acids with different number of carbon atoms arise from rumen-derived propionic acid rather than acetate as precursor for fatty acid synthesis, e.g. (C15 and C17). Branched chain fatty acids are derived from amino acids, valine, leucine and isoleucine (i.e. 4-methyl octanoic acid, C8:0 and 4-methyl nonanoic acid C9:0) and conjugated double bond fatty acids (*ie.* rather than being separated by CH_2 group, the bonds are on adjacent carbon atom). These wide ranges of products are a results of the rumen action that degrade plant structures and dietary fatty acids, some of which are incorporated into tissue lipids after its absorption in the small intestine. Conjugated linoleic acids (CLAS) are an important group of fatty acids in ruminants with 18 carbon and 2 conjugated double bonds. These have wide range of physiological action in the body of animals and also for meat consumers.

The saturated fatty acids in meat can be derived from the diet, produced in the rumen from unsaturated dietary fatty acids or synthesized from glucose or acetate in liver or adipose tissue.

The desaturase and elongase enzymes in ruminants can convert 18:2 to longer-chain n-6 fatty acids such as arachidonic acid (20:4n-6). Similarly, long chain n-3 fatty acids such as eicosapentaenoic acid (EPA; 20: 5N-3) and docosahexaenoic acid (DHA, 22: 6n-3) can be synthesised from the most common n-3 fatty acid -linoleinic acid (18: 3n-3) which is present in the leaves of plants and grasses.

Ruminants and non-ruminants species differ greatly in their proportions of PUFA in tissue and meat, whereas these are hardly changed by digestion in monogastric animals like pigs and poultry and are incorporated directly into adipose tissue. In ruminants they are extensively hydrogenated by micro-organisms in the rumen. The microbial action results in generally low levels (10% or less) of dietary PUFAS being available for absorption into body tissue after passing through rumen and a wide range of modified fatty acids. Thus ruminant fats have a major content of saturated fatty acids. The cold temperature performance of bio-fuel synthesised from these fats decreases with a rising content of saturated fatty acids.

Thus taking into consideration the cold temperature performance, monogastric animal fat like that of pig and poultry are preferred for the production of bio-diesel than that of ruminants like cattle, buffalo, sheep and goat. Moreover, pigs have the unique property of storing fat as a separate layer below the skin, which is easy to separate.

Table 8.1 Main reference parameters in the chemical analysis of some animal fat

Parameters		Unit	Poultry	Beef Tallow	Lard
Moisture	Max	%	0.5	0.50	0.50
Total fatty acids	Minim	%	90	94	94
Free fatty acids		%	8	4	4
Unsaponifiable matter	Max	%	1.0	1.5	1.5
Saponification value		%	190-196	190-202	190-194
Iodine value		g/100g	75-85	35-45	55-65
Peroxide value	Max	meq/1000g	10	10	10

Source: Baiao and Lara, 2005.

Bio-diesel Production

"The mono alkyl ester of long chain fatty acids derived from renewable lipid feed stock such as vegetable oil or animal fats, for use in compression ignition (diesel) engines and which meets the specification of ASTM D 6751 is known as bio-diesel " (National Bio-diesel Board, 2009).

The term "bio-diesel" is standardised as mono-alkyl (Methyl, propyl or ethyl) ester or Fatty Acid Methyl Ester (FAME).

Bio-diesel is a liquid which varies in colour between golden and dark brown depending on the production feedstock. It is immiscible with water, has a high boiling point and low vapour pressure. The flash point of bio-diesel (>130 °C) is significantly higher than that of petro diesel (64 °C). Bio-diesel has a density of "H" 0.88 g/cm3, less than that of water. Bio-diesel has no sulphur content.

Conversion of fatty acid in fats and oils to their simple alkyl esters known as bio-diesel produces a fuel that functions well in unmodified diesel engines, while substantially reducing the undesirable emission of un-burned hydrocarbons, particulates, carbon monoxide, polycyclic aromatic hydrocarbon and sulphur dioxide. In addition, bio-diesel reduced net emissions of carbon dioxide (carbon neutral), dependence on imported energy sources, increased agricultural income and fuel lubricity, thereby reducing ware in close tolerance fuel injection system.

Bio-diesel offers several fuel advantages over petroleum diesel including improved lubricity, a higher flash point, lower toxicity, bio-degradability and no net contribution to the greenhouse effect because it is made from renewable resources. In all diesel engines, it is advisable to use a blend of 20 per cent bio-diesel with 80 per cent petroleum diesel (B20). However, use of pure form of bio-diesel (B100) may require certain engine modifications to avoid performance problem and periodic maintenance.

Direct use and blending

Fat releases concentrated amounts of energy when burned, because of their chemical composition. However, their direct use in compression ignition engines was restricted due to high viscosity of 10 to 22 times higher than that of fossil diesel, which results in incomplete combustion and poor fuel atomisation and carbon deposits on the injector and valve seats causing serious engine fouling. Furthermore, acrolein (a highly toxic substance) is formed through thermal decomposition of glycerol. To overcome these constraints, three methods *viz.*, pyrolysis, micro-emulsification and transcsterification were developed. Among this, pyrolysis and micro-emulsification were discarded due to copper corrosion and injector needle sticking.

Transesterification

Transesterification (alcoholysis) is the general term used to describe the important class of organic chemical reactions where an ester is transformed into another by interchange of the alkoxy moiety. Transesterifcation involves stripping the glycerine from the fatty acids in the presence of a catalyst and replacing it with radicals from the anhydrous alcohol. It is a sequential three consecutive reversible reaction. In the first reaction, triglycerides are converted to diglycerides, then diglycerides to monoglycerides and in the last reaction monoglycerides are converted to glycerol. An ester is produced in each step and thus three ester molecules are produced from one molecule of triglyceride. A catalyst is usually used to enhance the reaction rate and bio-diesel yield. Since the reaction is reversible, excess alcohol is required to shift the equilibrium to the product side.

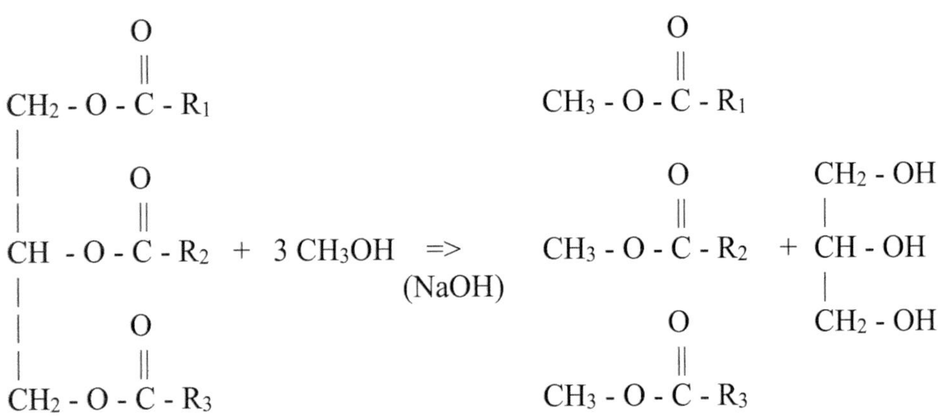

Triglyceride methanol mixture of fatty esters glycerin

Where R_1, R_2 and R_3 are long hydrocarbon chains sometimes called fatty acid chain.

Out of the above three methods, transesterification is the most viable process adopted, known so far for the lowering of fuel viscosity. It maintains the cetane number and heating value. It also gives glycerol as a by-product which has a commercial value.

There are three basic methods of ester production from oil / fat. They are 1. Base catalyzed transesterification 2. Acid catalyzed esterification and 3. Enzymatic catalysis. However, base catalyzed transesterification technique is the most commonly used method being the most economical process.

Base catalysed transesterification

The most common and traditional method of conversion of triglycerides found in fats and oils into fatty acid alkyl esters (otherwise known as bio-diesel) is by an alkali or base catalysed transesterification. The base catalysed reaction has a notably higher rate of reaction than the acid catalyst which can also be used for the transesterification process. One study suggested that base catalysed transesterification proceeds approximately 4000 times faster than acid catalysed transesterification. Common base catalysts used for transesterification include sodium hydroxide, sodium methoxide, and potassium hydroxide.

In the first step, the carbonyl carbon of the triglyceride molecule is attacked by the alkoxide ion to form a tetrahedral intermediate. Then this intermediate reacts with an alcohol to produce the alkoxide ion in the second step. The last step results in the rearrangement of the tertrahedral intermediate producing an ester and a diglyceride. This is repeated until three esters and glycerol are produced.

Acid catalyzed esterification

Esterification of Free Fatty Acid (FFA) is most effectively carried out using an acid catalyst, while a base catalyst is most effective for the transesterification of triglycerides. If diffusion is excluded, data have shown that the acid catalyzed esterification is of first order. It was generally observed that the highest rates of conversion were in the early stages of the reaction, usually in the first 30 mins in the total reaction period of 60 min. Sulphonic, sulfuric and hydrochloric acids are the most common acid catalysts used for esterification.

The conversion of FFA into fatty acid alkyl esters was most effective, when an acid catalysed esterification process was utilised. This process has high potential as the pretreatment step for reducing the FFA content of oil or fat, or as a technique for converting soap by-products into bio-diesel. High concentration of H_2SO_4 as catalyst could burn some of the oil which will then cause low yield of bio-diesel product.

Alternative processing techniques

Other alternative methods which are available for producing bio-diesel from animal fats and vegetable oils include the use of co-solvents, enzymes and supercritical methanol.

Kreutzer (1984) has described a high pressure and temperature (90 bar, 240°C) process which can transesterify the fats without prior removal or conversion of FFAs.

Co-solvents

Boocock *et al.* (1998) found that by using a nonreactive co-solvent such as tetrahydrofuran (THF) or methyl tertiary butyl ether, the reaction could proceed in a single phase, resulting in a much higher rate of transesterification. The results indicated that the use of a cosolvent could significantly reduce the temperature and reaction time necessary for transesterification of triglycerides. The hazards associated with THF (highly flammable, may generate explosive peroxides during storage if in contact with air), however, limits its attractiveness as a potential cosolvent.

Enzymes

Jackson and King (1996) reported a direct methanolysis of triglycerides using an immobilized lipase in flowing supercritical carbon dioxide. Corn oil was pumped in a carbon dioxide stream at a rate of four L/min and methanol at five L/min to yield more than 98 per cent fatty acid methyl esters. This process combined the extraction and transesterification of oil.

Nelson *et al.* (1996) used lipases *viz. Mucor miehei* (Lipozyme IM60), *Candida antarctica* (SP435), *Geotrichum candidum*, *Pseudomonas cepacia* (PS30), and *Rizopus delemar* as bio-catalysts for the conversion of fats and oils to bio-diesel. They carried out this investigation in order to overcome the processing problems encountered with the traditional chemical technique such as recovery of glycerol, amount of energy required, and issues with the conversion of high free fatty acid feed stocks.Though the bio-catalysts had the potential to improve some of the processing problems, they were very expensive which limited their use.

Supercritical methanol system

Super Critical Fluids (SCF) are produced by increasing both the temperature and pressure to reach a critical point at which gases and liquids are indinstinguishable fluids. At such super crtical conditions, Methanol, become an excellent solvent which has the capacity to dissolve the feedstock and readily reacts without catalysts.

Saka and Kusdiana (2001) claimed that it was possible to react the oil and methanol without a catalyst using super critical methanol, which eliminted the need for the water washing step. However, high temperture of 300 to 350 °C and methanol to oil molar ratios of 42:1 were required. Thus the energy cost of production were similar to that of catalytic production route.

Multiple step process

The requirement of catalysts varies considerably in the conversion of triglycerides and FFA to bio-diesel, though similar type of alcohols can be used in these techniques. Therefore, it was necessary to adopt a multiple step process for converting fats and oils containing both triglycerides and FFA to bio-diesel fuel.

For the effective convertion of low cost raw materials which contains high FFA's into bio-diesel, Canakci and Van Gerpen (2001) investigated an acid catalysed pre-treatment step followed by a base catalysed transesterification step. They found that by pre-processing the oil in two acid catalysed stages, with a water removal step in between, both the 20% and 40% palmitic acid samples could be successfully esterified down to an FFA content of less than 1%. They concluded that acid catalysed pretreatment step to convert the FFA to esters followed by an alkali catalysed step to convert the triglycerides should provide an effective and efficient method to convert high FFA feedstocks to bio-diesel.

By combining the benefits of both acid and base catalysts, feed stocks containing high concentrations of FFA can potentially be effectively converted to bio-diesel fuel. Thus for the effective conversion of rendered chicken oil to bio-diesel, John *et al.*(2014) reported that an acid catalysed esterification of FFA portion of the chicken fat could reduce the FFA from 7.5% to 0.70%, followed by the base catalysed transesterification of the triglycerides portion which could convert 97.62% of pretreated rendered chicken oil to bio-diesel.

Factors affecting bio-diesel yield in transesterification process

The bio-diesel yield is affected by different parameters such as moisture and FFA of oil, molar ratio of oil and alcohol, catalyst type and amount, reaction time, reaction temperature and stirring speeds.

Separation of methyl esters

Following the reaction, the glycerol is removed from the methyl esters, due to the low solubility of glycerol in the ester and due to its high molecular weight which causes it to settle down. This phase separation generally occurs quickly and is accomplished with a settling tank.

Bio-diesel washing

The water washing step is intended to remove any remaining catalyst, soap, salts, methanol or free glycerol from the bio-diesel. These contaminants have to be washed away as they will affect the quality of the bio-diesel and subsequently the engine performance.

Physico-chemical characteristics of Bio-diesel

The conversion of triglycerides into methyl esters through transesterification process reduced the molecular weight to one-third that of triglycerides, reduced the viscosity by a factor of about 87.5% per cent and increased the volatility marginally. These esters contain 10 to 11% oxygen by weight, which encourages more combustion than hydrocarbon-based diesel fuels in an engine. The cetane number of animal fat bio-diesel was 72.5. Bio-diesel had lower volumetric calorific values (about 8.75%) than diesel fuels but had a high cetane number and flash point. The cloud point and pour points of bio-diesel were 5.3 and 1.8 °C that where higher than those of diesel fuels.

Engine performance with Animal Fat Bio-diesel

In a test performed to assess the performance of a four cylinder CRDI engine using the bio-diesel blend B20 compared to that of commercial diesel (B00), the total fuel consumption was lower to that of commercial diesel in all range of the tested brake power. During all load conditions the bio-diesel blend (B20) showed a lower Brake Specific Fuel Consumption. While using B20, the mechanical efficiency was always higher than commercial diesel fuel. In all cases, the brake thermal efficiency was more than 3% for B20. At the maximum brake power of 58 kW, the smoke opacity of B20 was 47.14% less than that of diesel. While using B20 in the engine, the exhaust gas temperature was lower in the entire range of loads. But with increase in exhaust gas temperature, NO_x emission also increased.The better lubricating qualities of animal fat B20 prevented over heating of engine, which could prolong the engine life. The engine performances indicated that bio-diesel blend B20 improved total fuel consumption, brake specific fuel consumption, mechanical efficiency and brake thermal efficiency compared to commercial diesel. It also substantially reduced the engine emissions as proved by significantly lower smoke levels of 47.14 %. This is due to the high Cetane value of B20 which has shorter ignition delays, providing more time for fuel combustion, leading to more efficiency and less exhaust emission (John *et al.*, 2014).

Future Perspectives

Bio-diesel production technology provides opportunity to produce highly valued bio-fuel from dead animal, birds and slaughter waste, which can improve the engine efficiency and reduce vehicular pollution considerably. Simultaneously, this technology can also solve the major problem of unscientific disposal of dead birds and slaughter waste which is prevalent in developing countries, leading to severe environmental pollution. Kerala Veterinary and Animal Sciences University has established a pilot plant for producing bio-diesel from animal fats at the School of bio-energy and farm waste management (Fig. 8.1).

As such, animal production in the future may orient towards "Animals for food and fuel" which would revolutionise this sector leading to greater financial viability, production of renewable energy and environmental protection. Pigs and poultry would lead this sector because of their excellent attributes such as prolificacy, short generation interval and quick body weight gain for slaughter. Fat less pork at a premium price is gaining consumer acceptance phenomenally. This provides the separated fat to be converted to bio-diesel economically. Thus development and utilisation of new bio-mass feed stocks for the production of bio-fuel and development of next generation of more efficient bio-fuel conversion technologies awaits the future.

FIG 8.1 PILOT PLANT TO PRODUCE BIODIESEL FROM ANIMAL FATS

REFERENCE/SUGGESTED READING

Aherne, F.X., Hogberg, M.G., Kornegay, E.T and Shurson, G.C.1992. Management and Nutrition of the newly weaned pig. ***In Pork Industry Handbook*** (PIH-111).

Anon, 1996. Purdue University Cooperative Extension Service, West Lafayette, Pp1-4.

Appleby, M.C.,Pajor, E.A and Frazer, D.1991. Effect of management options on creep feeding by piglets. *Animal Production*.53.361-366.

Arora, R.L., Sukhdeo.,Chhabra, A.K., Satya Paul and Bist, G.S.1995. Studies of carcass traits in cross bred pigs. *Indian Veterinary Journal*. 71: 694-697.

Ayo, J.O. and Oladele, S.B. 1996. Transport stress in food animals. A review. *Niger. Vet.J. (Special edition)*, *1*: 58-68.

Baiao, N. C and L. J. C Lara, 2005. Oils and Fat in Broiler Nutrition. *Brazilian J. Poult. Sci.*, 7: 129-141.

Basic Animal Husbandry Statistics. 2018. Govt. of India, Dept. of A.H. and D. Ministry of Agriculture, New Delhi.1-74

Bhatia.S.S and Sharma P.K. 1979. Genetic studies on some economic traits in Landrace pigs. *Indian Veterinary Journal*. 56: 45-48.

Bhatia, S.S., Chhabra, A.K and Agarwal, D.K.1985. Effect of age at weaning on productivity of pigs. *Indian Journal of Animal Research*. 19: 48-50.

Biswas, D.K.1972. Studies on litter size and litter weight at birth and at weaning in Large White Yorkshire. *Indian Journal of Animal Research*. 22:5 142-145.

Biswas, M.2000. Animal husbandry development strategies for new millennium: *Poultry Advisor* 33 (3).

Blasco, A., Bidanel, J.P., Bolet, G., Haley, C and Santacreu, M.A.1993. The genetics of prenatal survival of pigs and rabbits: A review. *Livestock Production Sciences*. 37:1-21.

Boocock, D. G. B., S. Konar, V. Mao, C. Lee and S. Buligan, 1998. Fast Formation of High-Purity Methyl Esters from Vegetable Oils. *J. Am. Oil Chem. Soc.*, **75**: 198- 203.

Bowland, M.P., Evan A.C.O. Duffy, P., Hawken, P and Beard, A.P. 1986. Effect of progesterone treatment on oestrus synchronisation and fertility during the breeding season in sows. *Animal Reproduction*. 84: 349-358.

Burger, J.F.1952. Sex physiology of pigs. *Onderst. J.* Vet. Res.2:1-18.

Casey, D., Rathje, T and Johnson, R.1995. Index selection for components of litter size. *Nebraska Swine Report*. 42: 56-61.

Canakci, M and J. Van Gerpen. 2001. Biodiesel Production from Oils and Fats with High Free Fatty Acids. *Transaction of the Am SocAgric Eng.*, **44** (6): 1429-1436.

Chandrahas, Chhabra, A.K. Bhist. G.S & John Abraham.2004. Effect of three farrowing systems on the performance of cross bred primiparous sows and their litter. *Indian Journal of Animal Sciences*. 10 (6): 83-85.

Chandrahas, Chabra A.K and John Abraham.2006. Pig production under different housing systems:An economic analysis. *Indian Veterinary Journal.* 83: 299-302.

Chatterjee, J. K., Majumdar, S.C and Dasgupta, R.1988. Studies on some preweaning traits in Large White Yorkshire and crossbred pigs under intensive management. *Indian Veterinary Journal.* 65:683-686.

Chauhan, V.P.S., Arora, R.L., Chhabra, A.K and Bhat, P.N.1993. Litter traits and live weight of Landrace pigs in India. *Indian Veterinary Journal.* 70:29-31.

Chhabra, A.K., Bhatia, S.S., Sharma, N.K and Dutta, O.P.1990. Genetic study of litter traits in Large White Yorkshire pigs. *Indian Veterinary Journal.* 67:715-718.

Chhabra, A.K., SukhDeo, Raina, B.L and Agarwal, D.K. 1990.Effect of presence of boar on attainment of puberty in gilts. *Livestock Adviser.* 15:4-7.

Chauhan, V.P.S., Arora, R.L., Chhabra, A.K and Bhat, P.N.1993. Litter traits and live weights of Landrace pigs in India. *Indian. Vet. J.* 70: 32-34.

Clutton-Brock. T.H.1991. *The evolution of parental care.* Princeton University Press. Pp352.

Cole, D.J.A and Foxcroft, G.R. 1982. *Control of pig reproduction.* Butterworths London.328 Pp

Cole, D.J.A and Varley, M.A.2000. Boosting postweaning gains. *Pig farming*, 6 (2). 23-98 Pp 541-564.

Crighton, D.B., and Lamming, G.E. 1970. Endocrinology of anoestrum during laction in sow. *J. Endocri.*43: 507.

Delaware Bulletin 40. 2003. Delaware Department of Insurance. Pp.40.

Deo, S., Chhabra, A.K., Arora, R.L., Satya Paul and Bisht, G.S.1992. Genetic and non-genetic factors affecting growth performance in desi pigs and its crosses. *IndianJournal of Animal Sciences.* 62 (11): 1107-1108.

English, P.R., Smith, W.J and Maclean, A.1982. *The sow: Improving her efficiency*. 2nd Ed. Farming Press Limited. Pp-660.

English P.R and Morrison. V. 1984. Causes and prevention of pig mortality. *Pig News and Information, Commonwealth Agricultural Bureaux*. 5 (4): 369-376.

English, P.R.1991. Stockmanship, empathy and pig behaviour. *Pig Veterinary Journal*. 26.56-66.

FAO. 2007. *Food and Agricultural Organization production year book* – 2007.United Nations, Rome.

FAO. 2018. *Food and Agriculture Organization [FAO] Statistics*. United Nations.Available: http://faostat.fao.org/. [20 Feb. 2018].

Forbes. J.M. 1995. *Voluntary food intake and diet selection in farm animals*. CAB International. Wallingford Oxon. Uk. Pp.1-544.

Fahmy, M.H.1981. Factors influencing the weaning-to-oestrus interval in swine: A review. *World Rev. Anim. Prod.*17:15.

Fraser, D and Rushen, J. 1992. Colostrum intake by the new born piglets. *Canadian Journal of Animal Sciences*. 72. 1-13.

Gaur, G.K, Chhabra, A.K and Satya Paul. 1997. Growth intensity of Indigenous pigs from birth to slaughter age. *Indian Journal of Animal Sciences*. 53 (1): 97-100.

Goswami, R.N and Rraina,B.L. 1983. Body weight, daily weight gain and growth curve in Landrace pigs. *Indian Journal of Animal Sciences*. 10 (6) 83-85.

Grandin .T. 2012. *Livestock Handling and Transport*. 4th Edition. CAB International. Wallingford Oxon. UK. Pp-1-352.

Hausman, D.B., Hausman, G.J and Martin, R.J.1993. Endocrine regulation of foetal adipose tissue metabolism in the pig. *Biology of the Neonate*. 64, 116-126.

Hazel, L.N.1963. Performance of crossbred swine compared to purebred swine. *World conference on Animal Production*. European Asso. For Anim. Prod. Rome. 2: 189-198.

Hemsworth, P.H., Barnett, J.L and Hansen, C. 1981. The influence of handling by Humans on the behaviour, growth and corticosteroids in the juvenile female pig. ***Hormones and Behaviour***. 15. 396-403.

Hollis. G.R.1993. ***Growth of pigs***. CAB International. Wallingford Oxon. Uk. Pp-1-256.Serres. H. 1992. ***Manual of pig production in the tropics***. CAB International. Wallingford Oxon. Uk. Pp-1-288.

Houpt. K.A.2012. ***Domestic animal behavior for veterinarian and animal scientist***. 5th Edition. Wiley Blackwell. Pp-379.

Index of current research on pigs. 2015. (Number 41) Annual Publication. ISSN 0568- 2008.

Jackson, M. A and J. W. King, 1996. Methanolysis of seed oil in flowing super critical carbon dioxide. *J. Am. Oil Chem. Soc.*, **73**: 353-356.

Jarvis, S., Moinard, C., Robson, S.K., Sumner, B.E., Douglas, A.J., Seckl, J.R., Russell, J.A. and Lawrence, A.B., 2008. Effects of weaning age on the behavioural and neuroendocrine development of piglets. *Appl. Anim. BehaviourSci.110*(1): 166-181.

Jogi, S and Johar, K.S.1995. Selection indices for common Indian pigs. *Indian Veterinary Journal*.72. 252-253.

John Abraham.2003. Swine production in India: prospects and challenges. *Journal of* **Indian veterinary association**Vol; 1, Issue 1, 18-21.

John Abraham., Chhabra A.K and Chandrahas.2003. Effect of Split-weaning and Lighter-half Split-weaning along with Nursery feeding on the Reproductive performance of 'Landrace x Desi' Sows. *Indian Journal of Animal Sciences, 73(6):85-00.*

John Abraham., A. K. Chabra and Chandrahas (2004) Influence of different weaning methods on the survivability and economics of Landrace x Desi Pigs. *Indian Journal of Animal Production and Management* Vol. 20 (1-4) 1-7 December 2004.

John Abraham., Satyapaul, Chhabra, A.K and Chandrahas.2004.Effect of Early-weaning, Split-weaning and Nursery feeding programs on the growth of Landrace x Desi pigs. *Tropical Animal Health and Production,* 36 (6): 599-608.

John Abraham and Ramesh Saravanakumar (2012) Biodiesel production from animal fat an evergreen technology for the future energy security *Journal of Indian veterinary associationVol; 10, Issue 3, December 2012. 72-77.*

John Abraham.2014.*Practical aspects of establishing commercial farms.* Training Manual on Developing resource pool on entrepreneurial advisory services on livestock farming for the state of Kerala. 21 days training organised by the Directorate of Entrepreneurship, KVASU, Pookode. Pp 166-173.

John Abraham, Ramesh Saravana Kumar, Kulkarni V.V, Singh A. P and Visha. P.2014. Yield and quality Characteristics of rendered chicken oil. *Journal of American Oil Chemist Society* Vol. 91 (1) pp 133-141.

John Abraham and Ramesh Saravanakumar.2014.Scope of Producing Biodiesel from Poultry Fat. *Energy Blitz.* Vol-IV (111) June-July-2014; 19-23.

John Abraham, Ramesh Saravanakumar, Francis Xavier and Deepak Mathew. 2015. Biodiesel production from broiler chicken waste. International *Journal of Biological, Bio-molecular, Agricultural, Food and Biotechnological Engineering.* Vol: 9, No: 12, 2015.

Jyothi, S.B., John Abraham, Deepak, B. and Balusami, C. 2018. Bio-gas production potential of ruminant and non ruminant animal excreta. *Indian Journal of Animal Production and Management.* Vol. 33, 82-85.

Kelley, K.W. 1980. *Stress and Immune function*: A bibiliographic review. *Annales de Recherches Veterinarires.*11, 445-478.

Kotresh Prasad, Raseel, K and Sunanda, C.2016. A study of seasonal variations on productive performance of cross bred pigs in high range regions of Wayanad. *Int. J. Sci and Nat.*7(2) 2016: 387-389.

Kreutzer, U. R. 1984. Manufacture of fatty alcohols based on natural fats and oils. *J. Am. Oil Chem. Soc.*, **61**: 343- 348.

Kurian Thomas. 1980. Effect of environmental enrichment on the productivity of Large White Yorkshire Pigs. Personal communication.

Lakhani, G.P and Bhadoria, S.S.1989. Studies on preweaned economic traits in Large White Yorkshire. *Indian Journal of Animal Research*.23: 63-65.

Lakhani, G.P and Neema, R.K.1989. Studies on birth weight and weaning weight in Large White Yorkshire. *Indian Journal of Animal Research*.23: 13-15.

Lakhani, G.P., Kahlon, B.S., Jogi, S and Rao, M.L.V.1995. Behavioural pattern of sows during farrowing. *Indian Journal of Animal Research*. 29: 72-74.

Lakhani, G.P and Jogi, S. 1997. Ethogram of farrowing behavior in Desi pigs. *Indian Veterinary Medical Journal* 21:56-58.

Lakhani, G.P and Jogi, S. 1999. Genetic study of dressing percentage in desi pigs and their Large White Yorkshire grades. *Indian Journal of Animal Research*:33. 29-31.

Livestock Census. 2012. All India Report. Ministry of Agriculture, Department of Animal Husbandry, Dairying and Fisheries, Govt. of India, KrishiBhawan, New Delhi. Pp1-120.

Mahan, D.C and Lepine, A.J.1991. Effect of pig weaning weight and associated nursery feeding programs on subsequent performance to 105 kilograms body weight. *Journal of American Sciences*, 69.1370-1378.

Mahanta, S., Choudhury, B and Goswami, B.K.1986. Neonatal mortality in piglets. *Indian Journal of Animal Sciences*. 56:947-948.

Mahanta, S., Choudhury, B and Baruah, G.K.1987. Bacterial infection in neonatal piglets. *Livestock Adviser*. 12: 49-51.

Mount. L.E.1968. *The climatic physiology of the pig*. Edward Arnold. London.Pp. 322

Mukhopadhyay, A., Singh, R.L., Singh, S.K. Sharma, B.D. 1992. Factors affecting performance of pigs. *Indian J. Anim. Sci.* 61:438-442.

Mutton, W.J., 1987. The influence of photoperiod and light intensity on the farrowing sow and litter. *Group. 1*(2). p.235.

National Biodiesel Board. 2009. [Accessed on 23.11.2018] at http://www.biodiesel.org

Nelson, L. A., T. A. Foglia and W. N. Marmer, 1996. Lipase-Catalysed Production of Biodiesel. *J. Am. Oil Chem. Soc.*, **73**: 698- 703.

NRC.1998.*Nutrient requirements of Swine*. 10th ed. *National Acad. Press. Washington, DC.* Pp-237.

Ollivier, L and Sellier, P. 1982. Pig genetics: A review. *Annales de Genetiqueet de Selection Animale.* 14: 481-544.

Pajor, E.A., Frazer, D and Kramer, D.L.1991. Consumption of solid food by suckling pigs: individual variation and relation to weight gain. *Applied Animal Behaviour Sciences.* 32. 139-155.

Peters.2006. Vaccines for veterinary application. Buttorworth-Heinemann. London.Pp304.

Philips, P.A and Fraser, D.1993. Development in farrowing housing for sow and litter. *Pig News and Information*.14, 51N-55N.

Pond, W.G and Houpt, K.A.1978. *The Biology of the Pig*. Comstock Publishing. Ithaca.

Pourouchottamane.R., 2013.Effect of micro-environmental variations on biomolecular profile and performance of pig *J. Agrometeorology.* 15(1):*1-6*.

Pluske, J.R., Hampson, D.J and Williams, I.H.1997, Factors influencing the structure and function of the small intestine in the weaned pig;a review, *Livestock Production Reviews* **51**: 215-236.

Randall, G.C.B. 1992. Perinatal adaptation in animals. *Animal Reproduction Sciences.* 28: 309-318.

Ravi, A., Rao, D.S and Reddy. K.K.1999. Effect of sex on voluntary feed intake and growth performance in crossbred pigs. *Indian Journal of Animal Nutrition.* 16(1): 56-59.

Report.2001. Review committee report on All India Coordinated Research Project on Pigs. ICAR. New Delhi. 1-100.

Report. 2010. World Piggery Situation-2010.*Bulletin of the International Piggery Federation 446/2010.* Brussels. 446p.

Report. 2013. *Farm animal welfare compendium*: Statistics Pigs 1-13p.

Report. 2016. *Global Agriculture Information net work.* Report No. IN 6098.1-18.

Roppa. L. 2006. *"Global meat production: meeting the challenge in a changing world".* III Latin American Pork Congress. Brazil. Pp2-32.

Saka, S. K and D. Kusdiana, 2001. Biodiesel fuel from rapeseed oil as prepared in supercritical methanol, *Fuel.*,**80**: 225-231.

Singh, K.A., Nath, D.R and Sarkar, A.B. 1983. Certain economically important carcass Traits of castrated and uncastrated male Hampshire pigs. *Cherion*,12: 5-11.

Singh, K. I., Singh, R.L., Singh S.K., Sharma, B.D and Dubey, C.B.1990. Body weights and efficiency of feed utilisation in pigs. *Indian J. Anim.Sci.* 60:605-608.

Singh, K.P., Mishra, R.R and Singh R. 1992. Factors influencing preweaning growth of Landrace pigs in India. *Indian Journal of Animal Research.* 26: 81-84.

Sriraman, P.K and Reddy, K.K.1989a.Common diseases of pigs. *Livestock Adviser.* 14:5- 53.

Sriraman, P.K and Reddy, K.K.1989b.Common diseases of pigs. *Livestock Adviser.* 14:23-29.

Statistics.2015. *Pig Meat.* Danish Agriculture and Food Council. 1-48.

Shobhamani, B and Reddy, K.K.1997. Causes of preweaning mortality of Swine. *Indian Veterinary Journal.* 76: 25.

Tamuli, M.K., Rajkonwar, C.K., Goswami, B.K and Mahanta, S. 1986. Parturition behaviours in pig. *Indian Journal of Animal Reproduction.* 7 (2):73-84.

Varley, M.A. 1982. The time of weaning and its effects on reproductive function. *Control of Pig Reproduction* (ed. D.J.A. Cole and Fox Croft). Butter Worth London. Pp 459-478.

Varley. M.A. 1995. *The neonatal pig development and survival.*CAB International. Wallingford Oxon. Uk.. Pp-117-341.

Veum, T. and Odle.2001. Feeding Neonatal Pigs. In *Swine Nutrition.* 2nd Ed:Lewis A.J.Souther L.L.Eds CRC press LLC: NewYork, 671-690.

Warnick, A.C., Casida, L.E and Grummer, R.H.1950. The occurrence of oestrus and ovulation in post-partum sow. *J.Anim.Sci.,* 9:66-72.

Wathes, C.M and Charles D.R.1994. *Livestock Housing.* CAB International. Wallingford Oxon. Uk. Pp-1-450.

Weary, D.M., Jasper, J. and Hötzel, M.J. 2008. Understanding weaning distress. *Appl. Anim. Behaviour Sci.110*(1): 24-41.

Wegner, K., Lambertz, C., Das, G., Reiner, G. and Gauly, M., 2014. Climatic effects onsow fertility and piglet survival under influence of a moderate climate. *Animal, 8*(9): 1526-1533.

Williamson, G and. Payne W.J.A.1978. *An introduction to animal husbandry in the tropics* 3rded.London. Longman.

Young, L.D.1990. *Genetics of Swine*. Department of Agriculture, Agriculture Research Service, Clay Centre, Nabraska, Pp.70-76.

Yadav, B.L., Kumar, A and Singh, N.1991. Pig farming and breeding management for maximum production. *Livestock Advisor*. 16 (4): 15.

Zullo A., Hassan, C, Andraina A. 2006. *H. pylori infection during procedures.*Helicobactor pylori. Caister Academic Press. Pp 536.

INDEX

C

D

O

P

Q

T

U

V

W

For Product Safety Concerns and Information please contact our EU representative GPSR@taylorandfrancis.com Taylor & Francis Verlag GmbH, Kaufingerstraße 24, 80331 München, Germany